NIMS Monographs

Series editor

Naoki OHASHI

Editorial Board

Takahito OHMURA
Yoshitaka TATEYAMA
Takashi TANIGUCHI
Kazuya TERABE
Masanobu NAITO
Nobutaka HANAGATA
Kenjiro MIYANO

For further volumes:
http://www.springer.com/series/11599

The NIMS Monographs are published by the National Institute for Materials Science (NIMS), a leading public research institute in materials science in Japan, in collaboration with Springer. The series present research results achieved by NIMS researchers through their studies on materials science as well as current scientific and technological trends in those research fields.

These monographs provide readers up-to-date and comprehensive knowledge about fundamental theories and principles of materials science as well as practical technological knowledge about materials synthesis and applications.

With their practical case studies the monographs in this series will be particularly useful to newcomers to the field of materials science and to scientists and engineers working in universities, industrial research laboratories, and public research institutes. These monographs will be also available for textbooks for graduate students.

National Institute for Materials Science
http://www.nims.go.jp/

Shin-ichi Todoroki

Fiber Fuse

Light-Induced Continuous Breakdown of Silica Glass Optical Fiber

Shin-ichi Todoroki
National Institute for Materials Science
Tsukuba
Japan

ISSN 2197-8891   ISSN 2197-9502 (electronic)
ISBN 978-4-431-54576-7   ISBN 978-4-431-54577-4 (eBook)
DOI 10.1007/978-4-431-54577-4
Springer Tokyo Heidelberg New York Dordrecht London

Library of Congress Control Number: 2014935980

© National Institute for Materials Science, Japan. Published by Springer Japan 2014
This work is subject to copyright. All rights are reserved by the National Institute for Materials Science, Japan (NIMS), whether the whole or part of the material is concerned, specifically the rights of translation, reprinting, reuse of illustrations, recitation, broadcasting, reproduction on microfilms, or in any other physical way, and transmission or information storage and retrieval, electronic adaptation, computer software, or by similar or dissimilar methodology now known or hereafter developed. Exempted from this legal reservation are brief excerpts in connection with reviews or scholarly analysis or material supplied specifically for the purpose of being entered and executed on a computer system, for exclusive use by the purchaser of the work. Duplication of this publication or parts thereof is permitted only under the provisions of applicable copyright laws and applicable treaties, and permission for use must always be obtained from NIMS. Violations are liable to prosecution under the respective copyright laws and treaties.
The use of general descriptive names, registered names, trademarks, service marks, etc. in this publication does not imply, even in the absence of a specific statement, that such names are exempt from the relevant protective laws and regulations and therefore free for general use.
While the advice and information in this book are believed to be true and accurate at the date of publication, neither the authors nor the editors nor the publisher can accept any legal responsibility for any errors or omissions that may be made. NIMS and the publisher make no warranty, express or implied, with respect to the material contained herein.

Printed on acid-free paper

Springer is part of Springer Science+Business Media (www.springer.com)

Preface

This monograph deals with experimental aspects of the fiber fuse phenomenon. My research began 17 years after this phenomenon was discovered. At that time, it had begun to attract attention as a serious problem for the optical communication industry (see Fig. 1.2). Although my research results using an ultrahigh-speed camera provided us with new findings about this moving luminous object, I could not help feeling that more experimental facts were needed if we were to extract the underlying mechanism explaining this seemingly strange behavior. Since then, I have collected many photographs showing fiber fuse damage, and this collection has convinced me that the behavior is no longer strange.

This monograph starts with a chapter reviewing silica glass optical fibers and the fiber fuse phenomenon and continues with three subsequent chapters exploring the fiber fuse behavior in typical single mode fibers step-by-step. To assist the reader, 12 links to online video clips are provided in the text (see Box 0.1 in the next section as an example). The last chapter concludes the discussion from a practical point of view to encourage further research.

I hope you enjoy the process of solving this riddle and discover the beauty of the track left by a tiny comet running through a silica glass fiber.

Tokyo, January 2014   Shin-ichi Todoroki

Box 0.1 Portal site for fiber fuse research

- A full list of fiber fuse research papers and related information are available. Supplementary information related to this book can be found here.

http://fiberfuse.info
Moderator: S. Todoroki
Establishment: Feb. 8th, 2013

QR code of the above URL $\Longrightarrow$

Acknowledgments

I am deeply grateful to the following people who contributed their time and energy to this research.

Mr. Kazuhide Hanaka, Mr. Akira Sakamaki, Dr. Joji Kuwabara, Mr. Keisuke Aizawa, Mr. Yoshihiro Kondou, Mr. Arata Mihara, Mr. Yousuke Suzuki, and Mr. Yuhei Ueno (Photron Ltd.) for helping with the ultrahigh-speed videography experiments.

Dr. Mikiko Tanifuji, Mr. Kosuke Tanabe (National Institute for Materials Science), and Dr. Masao Takaku (Tsukuba University), for developing "NIMS eSciDoc," a digital library system including "imeji" suitable for self-archiving video clips.

Mr. Takashi Kobayashi, Mr. Hiroshi Ogino, Dr. Garcia Villora (National Institute for Materials Science), and Dr. Virginie Nazabal (Université de Rennes 1) for producing video clips for the general public.

Dr. Yoshito Shuto (Ofra Project) and Dr. Yosuke Mizuno (Tokyo Institute of Technology) for sharing interesting information.

Contents

Acronyms

DSF Dispersion-Shifted Fiber, an optical fiber whose dispersion is designed to be zero at 1,550 nm (the minimum-loss window of silica glass fibers).

FMF Few Mode Fiber, an optical fiber designed to carry few (more than one) light wave modes.

HAF Hole-Assisted Fiber, an optical fiber whose core is surrounded by a number of holes. See Fig. 1.10 (b-1).

MFD Mode Field Diameter, a structural parameter of single-mode fibers, namely the diameter at which the light intensity is $1/e^2$ of the maximum value.

NZDSF Nonzero Dispersion-Shifted Fiber, an optical fiber whose dispersion is designed to be negative at 1,550 nm.

SMF Single Mode Fiber, an optical fiber designed to carry only one light wave mode.

Chapter 1
Silica Glass Optical Fiber and Fiber Fuse

Не рассказывайте мне, что светит луна; Покажите мне отблеск ее света на разбитом стекле.

— Антон Павлович Чехов

Don't tell me the moon is shining; show me the glint of light on broken glass.

— Anton Pavlovich Chekhov

1.1 Introduction

A fiber fuse is a particularly eye-catching phenomenon that may appear doubly unexpected to a non-professional eye (see Box 1.1). Firstly, a bright spot runs along an optical fiber much more slower than expected. It looks like a tiny comet. Secondly, the spot runs toward the pump laser rather than running from it. These facts could be understood once we learned that a fiber fuse behaves like a grass fire (see Fig. 1.1). A solitary fire wave or a bright spot persists as an irreversible reaction region that is fueled from the front, emitting light and heat all around, and leaving cylinders or damage behind it.

Box 1.1 Videos providing macroscopic view of fiber fuse propagation.

(a) A brief description of the fiber fuse phenomenon for the general public. This was originally made as a public relations video for an open campus day at the National Institute for Materials Science, Japan.

http://www.youtube.com/watch?v=BVmIgaafERk
Duration: 1:43
Fiber: SMF-28e (Corning)
Pump laser: 1480 nm, 7 W
Speed: 30 fps

(b) Fiber fuse initiation and propagation.

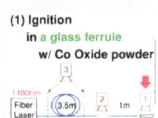
http://www.youtube.com/watch?v=yjX5dU1EkTk
Duration: 0:41
Fiber: SMF-28 (Corning)
Pump laser: 1480 nm, 9 W & 2 W
Speed: 30 fps
Reference: Fig. 1 in [62]

S. Todoroki, *Fiber Fuse*, NIMS Monographs, DOI: 10.1007/978-4-431-54577-4_1,
© National Institute for Materials Science, Japan. Published by Springer Japan 2014

Fig. 1.1 Comparison of
grass fire and a fiber fuse
with *arrows* showing the
energy flows into/from the
propagating reaction zone

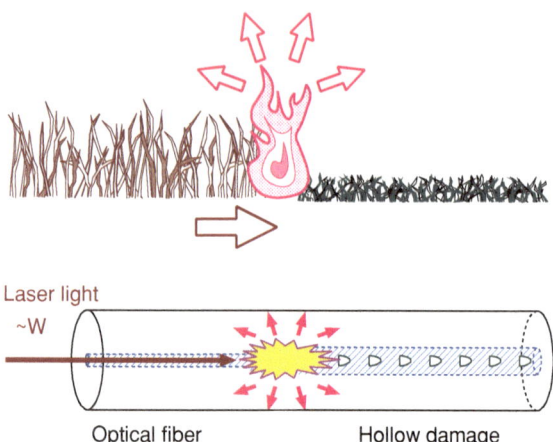

The name "fiber fuse" was given because of its striking resemblance to a fuse [29] or a detonating cord. Thus, we should be careful not to be confused with an electrical fuse or a device providing overcurrent protection.

Some scientists may wonder why the optical fiber absorbs the propagating light despite being made of silica glass, which is known to be ultimately transparent. This is because silica glass fiber becomes absorptive when it is heated above 1,050 °C [37] (see Sect. 1.2).

For the telecommunication engineers, this phenomenon is a pain in the neck because unintentional spot heating may cause an entire transmission line to break if the transmitted power exceeds the threshold for fiber fuse propagation, which is typically more than one watt for standard single-mode fibers (see Sect. 1.4). Thus, it imposes an unavoidable limit on the light power per fiber core that prevents the optical communication capacity from growing in the conventional way. Therefore, the current interest is to use special fibers that allow capacity growth without any increase in the power density [44]. Such approaches include the use of multicore fibers to increase the core numbers per unit cross section and the use of few-mode fibers with a larger core that reduces the power density.

However, this shift leaves a materials science problem undiscussed, namely what properties of silica glass are related to the behavior and stability of a fiber fuse. Unless we answer this question, the next generation system based on the above special fibers will face the same problem in the near future. This chapter discusses some key properties of silica glass that cause this phenomenon. Then, it describes the basic behavior of a fiber fuse and the safeguard technologies employed for current optical communication systems.

Fig. 1.2 Numbers of publications on the fiber fuse phenomenon indicated by total sum (*line*) and yearly increments (*bars*). *Source* http://fiberfuse.info/papers/ (see Box 0.1)

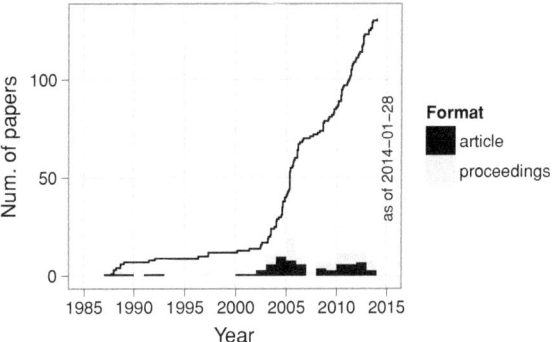

1.2 Silica Glass Optical Fibers

The invention of silica glass optical fiber with an attenuation of 20 dB/km at 632.8 nm in 1970 [36] was the starting point of the subsequent remarkable progress in optical communication that continues today [32]. It was triggered in 1966 by Kao [35], who won a Nobel Prize for physics in 2009. He predicted that a low-loss glass fiber waveguide of around 20 dB/km would be needed to realize a larger information capacity than the coaxial cable and radio systems available at that time.

Thanks to the feasibility of reducing impurities in silica glasses with the vapor-phase deposition technique, the attenuation loss of silica glass-based fibers was greatly reduced to 0.154 dB/km at 1.55 μm in 1986 [34], which is close to the theoretically predicted value. Two years later, the fiber fuse phenomenon was discovered [39], but it attracted limited attention because of the growing interest in optical communication. It was only at the beginning of the twenty-first century that the number of research papers began to increase (see Fig. 1.2) in response to the growing public interest in high-power laser technology. Finally, more than forty years after Kao's prediction, history is repeating itself. Now, the present fiber-based system is facing the problem of a capacity limit for data transmission as described above.

From the viewpoint of materials science, the interaction between a high-power laser beam and optical materials is one of the most important current topics. These materials should withstand exposure to an external laser beam. Silica glass optical fibers provide a good case study for this topic due to their high purity, simple structure, and wide proliferation at low cost.

The bright spot that we see in the fiber fuse phenomenon is high-density plasma (or optical discharge) confined in the fiber core and fed by the propagating light. Its temperature is experimentally estimated to be 10,000–5,000 K [19, 29], which is far larger than typical fiber drawing temperatures that are in excess of 2,000 °C [70]. A fiber with an outer diameter of 125 μm can withstand this heat because (1) the plasma is captured in a small core region with an outer diameter of about 10 μm and (2) its propagation speed is faster than the thermal conduction in silica glass (see Fig. 1.3).

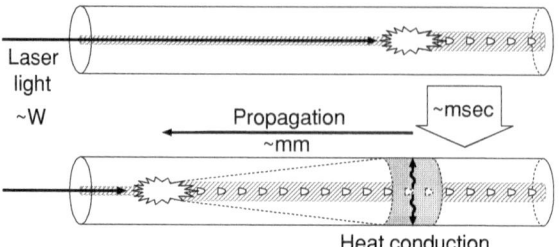

Fig. 1.3 Fiber fuse propagation and heat conduction

Table 1.1 Papers reporting a rough linear relationship between propagation speed and pump power

	Fiber type (Core diameter (μm))	Wavelength (nm)	Velocity (m/s)	Pump power (W)	Power density (MW/cm²)
Kashyap [37]	(5.40, 8.05)	514, 1,064	0.33–2.0	0.8–2.4	3.5–27
Hand et al. [29, 30]	(1.54)	514	1.9–3.2	0.38–0.54	20–29
Davis et al. [13]	SMF, DSF, etc.	1,064	0.37–3.0	0.5–4.75	3.1–17.6
Atkins et al. [9]	n.a.	1,480	0.32–0.69	–3.7	2.7–5.7
Bufetov and Dianov [10]	(4.7, 8.1, 9.5)	1.06×10^3	–5.9	–	1.2–92
Todoroki [62]	SMF	1,480	0.32–1.2	1.2 – 9.0	–
Dianov et al. [19]	(4.0, 5.75)	1.08×10^3	1–11	2–40	6–300
Lee et al. [43]	PM	1,480	0.3–0.51	1.6–3.2	3.7– 7.4
André [7]	SMF, DSF, NZDSF	1,480	0.4–0.96	2.0–3.5	2.3– 6.9
Abedin and Nakazawa [2]	SMF	1,480, 1,560	0.28–0.46	1.4–2.7	–
Domingues et al. [21]	G.657	1,480	0.51–0.68	2.5–4.0	–
Todoroki [68, 69]	SMF, NZDSF	1,310, 1,480	0.32–0.74	1.1–3.5	–

SMF standard single-mode fiber, *DSF* dispersion-shifted fiber, *PM* polarization-maintaining fiber, *NZDSF* non-zero DSF

Since the heat diffusion time through a 60-μm-thick glass layer is estimated to be a few milliseconds, which is roughly equal to the heat penetration time in fusion splicing [71], the fiber fuse travels a few millimeters during that period considering that a typical fusing speed is about 1 m per second (see Table 1.1).

Some optical fibers made of soft glasses are known to be damaged by a similar phenomenon, but its appearance is totally different. A thermally damaged region in chalcogenide and fluoride fibers spreads throughout the cross section and propagates very slowly (a few millimeters per second) without any hot plasma [17]. This is because of their poor thermal resistance. These glasses soften and devitrify at a few hundred degrees centigrade, which is too cool for plasma generation.

The confined plasma in a silica fiber originates from the local heating of a fiber delivering a few watts of light. As mentioned above, heated silica glass fiber absorbs light energy that generates additional heat in the core region and increases its temperature and absorbance. This feedback loop brings about a concentrated large temperature increase that generates high-density plasma in the fiber.

Fig. 1.4 Loss spectrum of silica glass [47], SiO [46], and silica glass optical fiber [33, 34]. A heat-induced loss increase is shown as a pair consisting of an *arrow* and an *open triangle* [37]

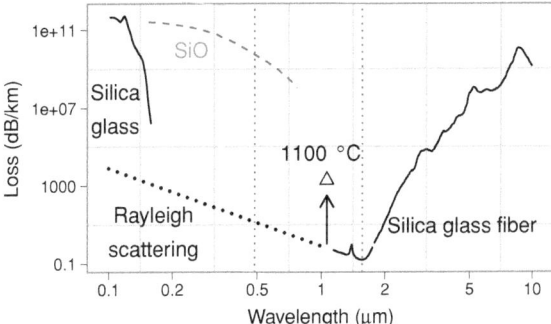

Kashyap provided experimental evidence of this heat-induced absorption in 1988 [37]. He found a steep increase in the loss of 1,064 nm light over 1,050 °C through a 1-m-long Ge-doped single-mode silica glass fiber. Figure 1.4 shows the wavelength dependence of the propagation loss through silica glass and fiber at room temperature where this heat-induced loss is shown as an open triangle. Later, Hand et al. [30] pointed out a similar behavior in the loss of 488 nm light in a multimode fiber. Recently, Dvoretskiy et al. [23] reported some loss increase data in the visible region at elevated temperatures to 1,200 °C in several silica fibers.

On the other hand, many researchers have reported fiber fuse propagation pumped at several wavelengths between 488 nm and 1.56 μm (see vertical dotted lines in Fig. 1.4). It is reasonable to expect that heat-induced absorption occurs in this wavelength range and that its origin is some thermally decomposed products of silica glass.

As the most probable origin among them, Shuto proposed SiO_x ($x \leq 1$) [54, 55], which is highly absorptive in the ultraviolet and visible region at room temperature [46] (see dashed curve in Fig. 1.4). Its near-infrared absorption is also known through the laser sintering of SiO powder at 1,064 nm [57]. Silicon monoxide is produced in silica glasses at elevated temperature through the following reaction [52],

$$SiO_2 \rightleftarrows SiO + \frac{1}{2}O_2. \tag{1.1}$$

This is suggested to occur during fiber fuse propagation by the fact that oxygen gas was detected in the hollow voids that remained after the passage of a fiber fuse [37].

Few experimental results have been published related to fiber fuse initiation. Davis et al. [13] reported that the threshold power density for initiation was of the order of 3 MW/cm^2 among typical single-mode fibers operated at 1,064 nm regardless of fiber type or core composition. They also made some interesting comments on initiation probability. Fuse initiation was harder for a pure silica core fiber than for fibers doped with other elements. This must be because doped elements with a low vapor pressure promote light-induced heat generation in the early stages of initiation. With a flame from a butane torch (∼1,700 K), the fuse initiated more easily when the flame was applied to the end of the fiber rather than to the middle. This is reasonable because the former geometry provides a direct heat flow to the core region.

With a similar geometry, fuse initiation was observed directly using an ultrahigh-speed camera [63, 64] (see Box 1.2). Figures 1.5 and 1.6 summarize this result. The heat source was a small amount of cobalt oxide powder, which was attached to the end surface of an optical fiber and converted the pump light energy (1,480 nm, 9 W) to heat and a visible emission.

Box 1.2 Video providing microscopic view of fiber fuse initiation.

• Original gray-scale image is converted to color-scale image.

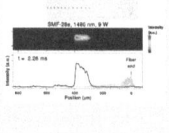

http://imeji.nims.go.jp/collection/8/item/12
Duration: 0:28
Fiber: SMF-28 (Corning)
Pump laser: 1480 nm, 9W
Camera: FASTCAM-APX RS (Photron)
Speed: 100,000 fps   Exposure time: 1 μs
References: Fig. 1.5, Fig. 1.6, and Fig. 3 in [63]

This video reveals the existence of a preliminary process that occurs before a fiber fuse begins to propagate stably at $t = 2.2$ ms. During this preliminary process, the extrinsic heat source boosts the intrinsic one, i.e., the decomposition products of silica glass. This reaction region is seen as a dark point that migrates slowly from the fiber end to a depth of 0.3 mm.

1.3 Fiber Fuse Propagation and Damage Left Behind

After the preliminary process, the confined plasma propagates at a constant speed. This is confirmed by some in situ observation results [61, 62] including Figs. 1.5 and 1.6. The speed increases roughly linearly with the pump power as many researchers have already reported (see Table 1.1). However, it is difficult to find a general relationship between different types of fibers even when the pump power is normalized as power density [13]. Recently, an improvement in the accuracy of velocity measurement revealed a deviation from the linearity [68, 69] (see Sect. 2.4).

These data are useful for validating the different fiber fuse heat flow models proposed by several researchers and listed in Table 1.2. In these calculations, a fiber fuse is treated as a point or a cylinder with no internal structure at which some of the following three heat processes are considered, i.e., light-induced heat generation, thermal diffusion, and radiation. Akhmediev et al. [5] demonstrated that a fiber fuse can be described as a one-dimensional dissipative soliton, a solitary wave of a reaction zone, that exists only in the presence of an external energy supply and internal loss in an open system far from equilibrium [4] like a grass fire (see Fig 1.1). As shown in

Fig. 1.5 In situ images of fiber fuse initiation captured by the ultrahigh-speed video camera (see Box 1.2). Visible light emission was recorded as a grayscale image and converted to color-scale image. The fiber end is located at $x = 0$. See also Fig. 1.6

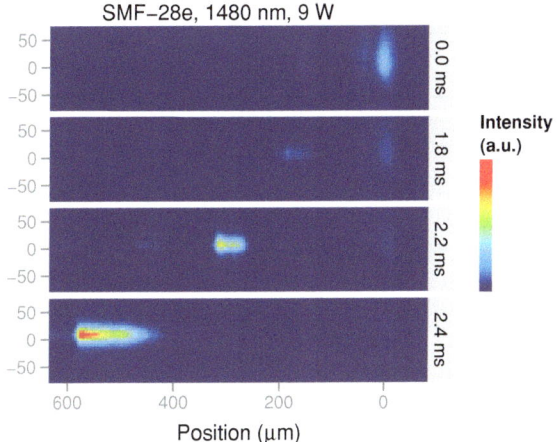

Fig. 1.6 a Intensity profiles along the center of the snapshots shown in Fig. 1.5 and **b** a photograph of the corresponding damage [63]. The laser pumping started several seconds before $t = 0.0$ ms

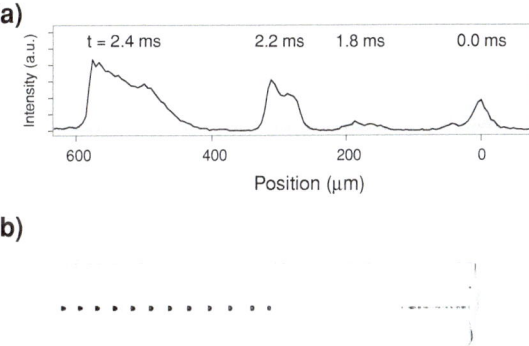

Fig. 1.7, a dissipative soliton is different from a classical soliton that transfers energy and/or mass.

After the passage of a fiber fuse, a hollow void train is left behind along the core region. Figure 1.8 shows the variation in the void shape left in a typical single-mode fiber pumped with a 1,480 nm laser light whose power was reduced from 7 W (e) to 1.2 W (a). After the initiation, the voids look like bullets placed at regular intervals pointing in the opposite direction to the fuse propagation ((e) and (d)). The interval between the bullet-like voids is reported to vary roughly linearly with the pump power as shown in Table 1.3. Then, the periodicity is lost (c), appears once more with smaller voids (d), and is broken again toward the termination point (a).

A pair of horizontal lines surrounds these voids in these photographs. This is the border surface of the region modified by the passage of the hot plasma. Their diameters are larger than the original core size, and they decrease as the pump power is reduced. In order to deepen our knowledge about this modified area, it is worth

Table 1.2 Numerical calculations of fiber fuse propagation

	Method	Source of fiber fuse velocity data
Hand et al. [29]	NSHCE	Hand et al. [29]
Kashyap [38], Kashyap et al. [40]	FEM	Kashyap [37]
Shuto et al. [55, 56]	NSHCE, FDM	Atkins et al. [9], Davis et al. [13], Kashyap [37]
Bumarin and Yakovlenko [11], Golyatina et al. [25]	NSHCE	Bufetov and Dianov [10], Todoroki [62]
Akhmediev et al. [5], Ankiewicz et al. [8]	FDM (dissipative soliton model)	
Facão [24], Rocha [50]	FDM	André [7]
Gorbachenko [26]	FDM	Bufetov and Dianov [10]

NSHCE non-stationary heat conduction equation, *FEM* finite element method, *FDM* finite difference method

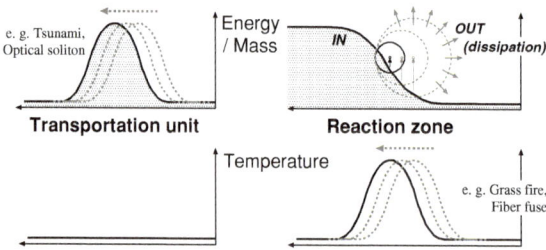

Fig. 1.7 Illustrations showing classical soliton (*left*) and dissipative soliton (*right*, see also Fig. 1.3)

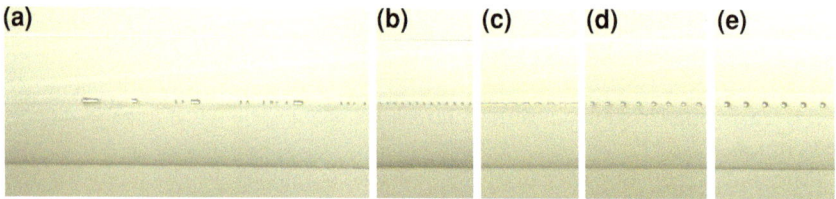

Fig. 1.8 A fiber fuse damage train in a single-mode fiber (Corning SMF-28). After the initiation of the fiber fuse, the pump laser power (1,480 nm) decreased from 7 W (**e**), 3.5 W (**d**) to 1.2 W (**a**) until its self-extinction at the end of the void train [66]

referring to the papers on bulk silica glass modification with a continuous-wave laser briefly summarized in Appendix A.

Table 1.3 Papers reporting the pump power dependence of the periodic void interval (Wavelength: 1,480 nm)

	Fiber type	Interval (μm)	Pump power (W)	Power density (MW/cm^2)
Atkins et al. [9]	n.a.	$1.5d_0 - 1.7d_0$	–3.7	2.7–5.7
Todoroki [62], Fig. 3.1	SMF	12.7–22.1	2.0–9.0	–
André [7]	SMF, DSF, NZDSF	10.5–14.7	2.0–3.5	2.3–6.9
Domingues et al. [21]	G.657	11.8–13.7	2.5–4.0	–

d_0 core diameter

Table 1.4 Papers reporting threshold power for fiber fuse propagation

	Fiber type (Core diameter (μm))	Wavelength (nm)			Available plot vs MFD (unit or slope)
Kashyap [37]	(5.40, 8.05)	1,064			
Dianov et al. [18]	SMF-28, etc.	1.06, 1.21, 1.48 × 10^3			I_{th} (MW/cm^2)
Nishimura et al. [45], Seo et al. [53]	SMF, DSF	1,064, 1,467			P_{th} (0.15 W/μm)
Takenaga et al. [59]	SMF, HAF, etc.			1,550	P_{th} (0.14 W/μm)
Takara et al. [58]	SMF, HAF, etc.		1,480		P_{th} (W)
Abedin and Nakazawa [2]	SMF-28		1,480,	1,560	
Rocha et al. [49]	SMF-28, DSF, NZDSF		1,480		
Todoroki [65]	SMF-28e	1,310,	1,480		
Todoroki [69]	NZDSF	1,310,	1,480		

MFD mode field diameter, *HAF* hole-assisted fiber

Chapters 3 and 4 discusses the formation mechanisms of these void patterns, including some irregular void patterns reported as long non-periodic filaments [12, 20, 29] and quasi-periodic patterns [20, 22].

1.4 Safeguard Technologies

Once a fiber fuse is generated, it continues to propagate unless the pump power supply is intercepted or is outrun by its energy dissipation. Therefore, the fundamental principle as regards preventing fiber fuse damage is to avoid delivering a light that is strong enough to allow fiber fuse propagation. The threshold power for its propagation is known to vary with fiber geometry. Many researchers have published its values under various conditions (see Table 1.4) and found that it increases linearly with mode field diameter (MFD) [18, 53].

Figure 1.9 shows threshold power values for commercial single-mode fibers, SMF-28 and SMF-28e (Corning). The linear dependence on pump wavelength is mainly due to the variation in MFD. In addition, it is expected to include the wavelength dependence of the photothermal conversion efficiency at elevated temperature, which has never before been evaluated.

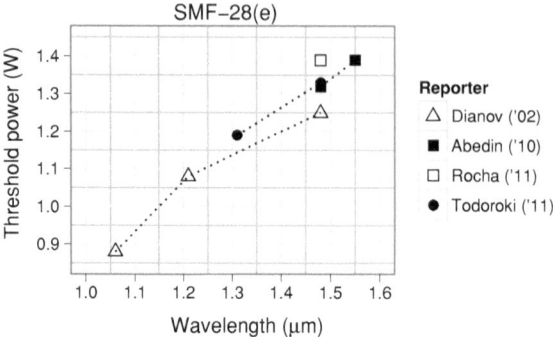

Fig. 1.9 Wavelength dependence of threshold power for fiber fuse propagation through SMF-28 and SMF-28e (Corning). *Source* [3, 18, 49, 65]

The threshold values at 1,480 nm in Fig. 1.9 are widely dispersed for the following two reasons. Firstly, the MFD may vary sample by sample within the product specifications, for example, 10.4 ± 0.5 μm at 1,550 nm. Considering the proportionality constant of 0.14 W/μm listed in Table 1.4, the variation of the threshold power is expected to be ± 0.07 W. Secondly, threshold power measurements are sensitive to the speed of the pump power reduction rate but no common procedure can be found in the literature. For example, one definition is the value at which a fiber fuse disappears during pump power reduction and another is the lowest power at which fiber fuse propagation occurs [65, 67] where the power reduction rate is zero.

To deliver light at beyond the threshold power, we must have some safeguard mechanisms for fiber fuse termination. Two technologies have been proposed: in-line termination devices and remote detection. The former is suitable for newly installed transmission lines while the latter is applicable for existing cables.

A fiber fuse can be terminated spontaneously at a modified segment in which the plasma becomes deactivated by one of the following two mechanisms. The first is that the pump light is deconcentrated so that it feeds less energy to the plasma. This is realized by local modification of the waveguide structure to expand the MFD of the propagating light (see Fig. 1.10a), i.e., inserting a segment with a larger MFD (a-1), a heat-treated segment with an expanded core (a-2) or a microwire (a-3). The other mechanism is one where the plasma and surrounding melt are exposed to the empty space near the core (see Fig. 1.10b). The segments with vacancies include hole-assisted fiber (b-1), photonic crystal fiber (b-2), and a cladding layer thinned by chemical etching (b-3). It should be noted that the empty space also modifies the MFD. Related papers are listed in Table 1.5.

The other safeguard technology is fiber fuse remote detection, which triggers the shutdown of the light source thus terminating the fiber fuse. A fiber fuse is detected by the reflection of a probe light at the fiber fuse (see Fig. 1.11a) or by a sensor made of another optical fiber that is thermally coupled with the main fiber and that responds to a heat pulse via the thermal expansion of a fiber Bragg grating (FBG,

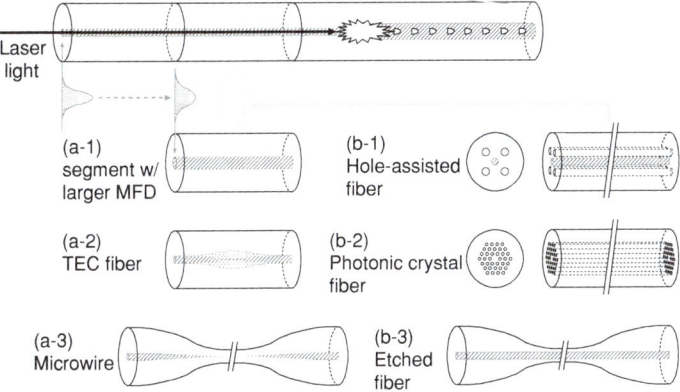

Fig. 1.10 Structures of fiber fuse terminators. See also Table 1.5

Table 1.5 Papers reporting fiber fuse terminators

Mechanism	Traps shown in Fig. 1.10	Reference and patents
Pump power density reduction	(a-2) Thermally expanded core	Hand and Birks [28]
	(a-1) MFD-expanded segment	Wyatt (2001) US 7162161
	(a-2) Thermally expanded core	Yanagi et al. [72], JP2002-372636-A
	(a-3) Microwire	Rocha et al. [51]
Empty space exposure	(b-3) Cladding thinned by etching	Dianov et al. [14, 15], EP 1528696
	(b-1) Hole-assisted fiber	Takenaga et al. [60], US 8244091
	(b-2) Photonic crystal fiber	Hanzawa et al. [31]
	Hollow optical fiber	Ha et al. [27]

Fig. 1.11 Two mechanisms of fiber fuse remote detection. See also Table 1.6

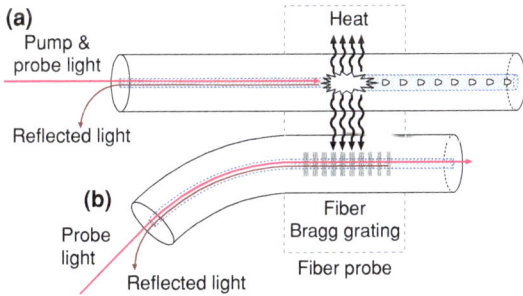

see Fig. 1.11b). The former signal is modulated by the dynamic structural change of the fiber fuse in the radio frequency range including periodic void formation and a Doppler shift [1]. The former detection is realized simply by using a commercial optical time-domain reflectometer (OTDR) [2]. In the latter system, a heat pulse of a few degrees centigrade was detected as an increase in the Bragg wavelength. Recently, Kinoshita et al. [41] demonstrated a detection and termination system using a Si photodiode (PD) for monitoring the visible light from a fiber fuse.

Table 1.6 Papers reporting fiber fuse remote detection

Detection	Sensor	Reference
(a) Back-reflected light at fiber fuse	PD, OTDR	Abedin and Morioka [1], Abedin and Nakazawa [3], Abedin et al. [2]
(b) Heat pulse	FBG	Rocha et al. [48]
(c) Visible emission	PD	Kinoshita et al. [41]

Another approach for avoiding a fiber fuse is to replace the entire optical line with a special fiber with a higher threshold power for fiber fuse propagation, i.e., various holey fibers including photonic crystal fibers and hole-assisted fibers. Some of them are reported to exhibit a threshold power at least 10 times higher than that of standard single-mode fiber [16, 31, 42].

1.5 Conclusion

A fiber fuse appears as a result of the heat-induced absorption of silica glass optical fiber pumped with a light strong enough to generate plasma that is confined in the core region and moves faster than the heat diffuses in the glass. It behaves like a grass fire and leaves a train of hollow voids. The telecommunication industry has been developing various safeguard technologies against this phenomenon, whereas its basic nature is still unclear from viewpoint of materials science.

See also Prof. Kashyap's survey [38] and two other review articles [6, 66].

References

1. K.S. Abedin, T. Morioka, Remote detection of fiber fuse propagating in optical fibers, in *Proceedings of Optical Fiber Communication/National Fiber Optic Engineers Conference* (2009). doi:10.1364/OFC.2009.OThD5 (OThD5)
2. K.S. Abedin, M. Nakazawa, Real time monitoring of a fiber fuse using an optical time-domain reflectometer. Opt. Expr. **18**(20), 21315–21321 (2010). doi:10.1364/OE.18.021315
3. K.S. Abedin, M. Nakazawa, T. Miyazaki, Backreflected radiation due to a propagating fiber fuse. Opt. Expr. **17**(8), 6525–6531 (2009). doi:10.1364/OE.17.006525
4. N. Akhmediev, A. Ankiewicz, Dissipative solitons in the complex Ginzburg-Landau and Swift-Hohenberg equations, in *Dissipative Solitons*, Lecture Notes in Physics, ed. by N. Akhmediev, A. Ankiewicz (Springer, Berlin, 2005), pp. 1–17, doi:10.1007/10928028_1 (ISBN 978-3-540-31528-5)
5. N. Akhmediev, J. St, P. Russell, M. Taki, J.M. Soto-Crespo, Heat dissipative solitons in optical fibers. Phys. Lett. A **372**(9), 1531–1534 (2008). doi:10.1016/j.physleta.2007.09.049
6. P. André, A. Rocha, F. Domingues, M. Facão, Thermal effects in optical fibres, in *Developments in Heat Transfer*, ed. by M. A. dos Santos Bernardes (InTech. Croatia, 2011), pp.1–20, doi:10.5772/22812 (ISBN 978-953-307-569-3)
7. P.S. André, M. Facão, A.M. Rocha, P. Antunes, A. Martins, Evaluation of the fuse effect propagation in networks infrastructures with different types of fibers, in *Proceedings of Opti-*

cal Fiber Communication/National Fiber Optic Engineers Conference (2010). doi:10.1364/NFOEC.2010.JWA10 (JWA10)

8. A. Ankiewicz, W. Chen, J. St, P. Russell, M. Taki, N. Akhmediev, Velocity of heat dissipative solitons in optical fibers. Opt. Lett. **33**(19), 2176–2178 (2008). doi:10.1364/OL.33.002176

9. R.M. Atkins, P.G. Simpkins, A.D. Yablon, Track of a fiber fuse: a rayleigh instability in optical waveguides. Opt. Lett. **28**(12), 974–976 (2003). doi:10.1364/OL.28.000974

10. I.A. Bufetov, E.M. Dianov, Optical discharge in optical fibers. Physics-Uspekhi **48**(1), 91–94 (2005). doi:10.1070/PU2005v048n01ABEH002081

11. E.D. Bumarin, S.I. Yakovlenko, Temperature distribution in the bright spot of the optical discharge in an optical fiber. Laser Phys. **16**(8), 1235–1241 (2006). doi:10.1134/S1054660X06080123

12. D.D., Davis, S.C. Mettler, D.J. DiGiovani, Experimental data on the fiber fuse, ed. by H.E. Bennett, A.H. Guenther, M.R. Kozlowski, B.E. Newnam, M.J. Soileau, in *27th Annual Boulder Damage Symposium: Laser-Induced Damage in Optical Materials: 1995, SPIE Proceedings*, vol. 2714, Boulder, CO, USA, pp. 202–210 (1996). 30 Oct 1995, doi:10.1117/12.240382

13. D.D. Davis, S.C. Mettler, D.J. DiGiovani, A comparative evaluation of fiber fuse models, ed. by H.E. Bennett, A.H. Guenther, M.R. Kozlowski, B.E. Newnam, M.J. Soileau, in *Laser-Induced Damage in Optical Materials: 1996, SPIE Proceedings*, vol. 2966, Boulder, CO, USA, pp. 592–606 (1997). 7 Oct 1996, doi:10.1117/12.274220

14. E.M. Dianov, I.A. Bufetov, A.A. Frolov, Destruction of silica fiber cladding by the fuse effect, in *OFC 2004 Technical Digest*. Los Angels (2004) (TuB4)

15. E.M. Dianov, I.A. Bufetov, A.A. Frolov, Destruction of silica fiber cladding by the fuse effect. Opt. Lett. **29**(16), 1852–1854 (2004b). doi:10.1364/OL.29.001852

16. E.M. Dianov, I.A. Bufetov, A.A. Frolov, Y.K. Chamorovsky, G.A. Ivanov, I.L. Vorobjev, Fiber fuse effect in microstructured fibers. IEEE Photon. Technol. Lett. **16**(1), 180–181 (2004c). doi:10.1109/LPT.2003.820465

17. E.M. Dianov, I.A. Bufetov, A.A. Frolov, V.M. Mashinskii, V.G. Plotnichenko, M.F. Churbanov, G.E. Snopatin, Catastrophic destruction of fluoride and chalcogenide optical fibers. Electron. Lett. **38**(15), 783–784 (2002a). doi:10.1049/el:20020539

18. E.M. Dianov, I.A. Bufetov, A.A. Frolov, V.G. Plotnichenko, V.M. Mashinskii, M.F. Churbanov, G.E. Snopatin, Catastrophic destruction of optical fibres of various composition caused by laser radiation. Quantum Electron. **32**(6), 476–478 (2002b). doi:10.1070/QE2002v032n06ABEH002226

19. E.M. Dianov, V.E. Fortov, I.A. Bufetov, V.P. Efremov, A.E. Rakitin, M.A. Melkumov, M.I. Kulish, A.A. Frolov, High-speed photography, spectra, and temperature of optical discharge in silica-based fibers. IEEE Photon. Technol. Lett. **18**(6), 752–754 (2006). doi:10.1109/LPT.2006.871110

20. E.M. Dianov, V.M. Mashinskii, V.A. Myzina, Y.S. Sidorin, A.M. Streltsov, A.V. Chickolini, Change of refractive index profile in the process of laser-induced fiber damage. Sov. Lightwave Commun. **2**, 293–299 (1992)

21. F. Domingues, A. Rocha, P. Antunes, A.R. Frias, R.A.S. Ferreira, P.S. André, Evaluation of the fuse effect propagation velocity in bend loss insensitive fibers, in *Technical Digest-17th OptoElectronics and Communications Conference, OECC2012*, pp. 799–800 (2012). doi:10.1109/OECC.2012.6276636 (6C3-2)

22. T.J. Driscoll, J.M. Calo, N.M. Lawandy, Explaining the optical fuse. Opt. Lett. **16**(13), 1046–1048 (1991). doi:10.1364/OL.16.001046

23. D.A. Dvoretskiy, V.F. Hopin, A.N. Gur'yanov, L.K. Denisov, L.D. Ishakova, I.A. Bufetov, Optical losses in silica based fibers within the temperature range from 300 to 1500 K. Sci Ed. Electron. Sci. Tech. J. 5 (2013). doi:10.7463/0513.0554843 (in Russian)

24. M. Facão, A.M. Rocha, P.S. André, Traveling solutions of the fuse effect in optical fibers. J. Lightwave Technol. **29**(1), 109–114 (2011). doi:10.1109/JLT.2010.2094602

25. R.I. Golyatina, A.N. Tkachev, S.I. Yakovlenko, Calculation of velocity and threshold for a thermal wave of laser radiation absorption in a fiber optic waveguide based on the two-dimensional nonstationary heat conduction equation. Laser Phys. **14**(11), 1429–1433 (2004)

26. V.J. Gorbachenko, A.Y. Dovzhenko, A.G. Merzhanov, E.N. Rumanov, V.E. Fortov, O.E. Yachmeneva, Propagation limits for a slow wave of optical breakdown in a fiber light guide. Dokl. Phys. **55**(8), 384–387 (2010)
27. W. Ha, Y. Jeong, K. Oh, Fiber fuse effect in hollow optical fibers. Opt. Lett. **36**(9), 1536–1538 (2011). doi:10.1364/OL.36.001536
28. D.P. Hand, T.A. Birks, Single-mode tapers as 'fibre fuse' damage circuit-breakers. Electron. Lett. **25**(1), 33–34 (1989). doi:10.1049/el:19890024
29. D.P. Hand, J. St, P. Russell, Solitary thermal shock waves and optical damage in optical fibers: the fiber fuse. Opt. Lett. **13**(9), 767–769 (1988). doi:10.1364/OL.13.000767
30. D.P. Hand, J.E. Townsend, P.S.J. Russell, Optical damage in fibres: the fibre fuse, in *Digest of Conference on Lasers and Electro-Optics*, Anaheim, US, Paper WJ1 (1988)
31. N. Hanzawa, K. Kurokawa, K. Tsujikawa, T. Matsui, K. Nakajima, S. Tomita, M. Tsubokawa, Suppression of fiber fuse propagation in hole assisted fiber and photonic crystal fiber. J. Light-wave Technol. **28**(15), 2115–2120 (2010). doi:10.1109/JLT.2010.2052913
32. J. Hecht, City of Light: The Story of Fiber Optics (Oxford University Press, Oxford, 2004). (Revised & expanded paperback edition) (ISBN 978-0195162554)
33. T. Izawa, S. Sudo, *Optical Fibers: Materials and Fabrication* (KTK Scientific Publishers, Tokyo, 1987) (ISBN 978-9027723789)
34. H. Kanamori, H. Yokota, G. Tanaka, M. Watanabe, Y. Ishiguro, I. Yoshida, T. Kakii, S. Itoh, Y. Asano, S. Tanaka, Transmission characteristics and reliability of pure-silica-core single-mode fibers. J. Lightwave Technol. **4**(8), 1144–1150 (1986). doi:10.1109/JLT.1986.1074837
35. K.C. Kao, G.A. Hockham, Dielectric-fibre surface waveguide for optical frequencies. Proc. Inst. Electr. Eng. **113**(7), 1151–1158 (1966). doi:10.1049/piee.1966.0189
36. F.P. Kapron, D.B. Keck, R.D. Maurer, Radiation losses in glass optical waveguides. Appl. Phys. Lett. **17**(10), 423–425 (1970). doi:10.1063/1.1653255
37. R. Kashyap, Self-propelled self-focusing damage in optical fibres, in *Lasers '87: Proceedings of the 10th International Conference on Lasers and Applications*, Lake Tahoe, NV, 7–11 Dec 1987, pp. 859–866. STS Press, McLean (1988)
38. R. Kashyap, Fiber fuse - from a curious effect to a critical issue: a 25th year retrospective. Opt. Expr. (2013). doi:10.1364/OE.21.006422
39. R. Kashyap, K.J. Blow, Observation of catastrophic self-propelled self-focusing in optical fibres. Electron. Lett. **24**(1), 47–49 (1988). doi:10.1049/el:19880032
40. R. Kashyap, A. Sayles, G.F. Cornwell, Heat flow modeling and visualization of catastrophic selfpropagating damage in singlemode optical fibers at low powers, ed. by H.E. Bennett, A.H. Guenther, M.R. Kozlowski, B.E. Newnam, M.J. Soileau, in *Laser-Induced Damage in Optical Materials: 1996, SPIE Proceedings*, vol. 2966, Boulder, CO, USA, 7 Oct 1996, pp. 586–591 (1997). doi:10.1117/12.274219
41. T. Kinoshita, N. Sato, M. Yamada, Detection and termination system for optical fiber fuse, in *OptoElectronics and Communications Conference Held Jointly with 2013 International Conference on Photonics in Switching (OECC/PS)* (2013) (Paper WS4-6)
42. K. Kurokawa, Optical fiber for high-power optical communication. Crystals **2**(4), 1382–1392 (2012). doi:10.3390/cryst2041382
43. M.M. Lee, J.M. Roth, T.G. Ulmer, C.V. Cryan, The fiber fuse phenomenon in polarization-maintaining fibers at 1.55 μm, in *Proceedings of the Conference on Lasers and Electro-Optics (CLEO)* (2006) (JWB66)
44. T. Morioka, New generation optical infrastructure technologies : "EXAT initiative" towards 2020 and beyond, in *Technical Digest-14th OptoElectronics and Communications Conference, OECC2009*, p. FT4 (2009). doi:10.1109/OECC.2009.5213198
45. N. Nishimura, K. Seo, M. Shiino, R. Yuguchi, Study of high-power endurance characteristics in optical fiber link, in *Technical Digest of Optical Amplifiers and Their Applications*, pp. 193–195 (2003) (TuC4) (We.P.20)
46. H.R. Philipp, Optical properties of non-crystalline Si, SiO, SiO_x and SiO_2. J. Phys. Chem. Solids **32**(8), 1935–1945 (1971). doi:10.1016/S0022-3697(71)80159-2

47. H.R. Philipp, Silicon dioxide (SiO$_2$) (glass), in *Handbook of Optical Constants of Solids*, ed. by E.D. Palik (Academic Press, New York, 1985), pp. 749–763 (ISBN 978-0125444200)

48. A.M. Rocha, P. Antunes, F. Domingues, M. Facão, P.S. André, Configuration for detecting the fiber fuse propagation using a FBG sensor, in *12th International Conference on Transparent Networks*. Munich, Germany (2010). doi:10.1109/ICTON.2010.5549119 (We.P.20)

49. A.M. Rocha, F. Domingues, M. Facão, P.S. André, Threshold power of fiber fuse effect for different types of optical fiber, in *The 13th International Conference on Transparent Optical Networks (ICTON 2011)*, pp. 1457–1549, Stockholm, Sweden (2011). doi:10.1109/ICTON.2011.5971025 (Tu.P.13)

50. A.M. Rocha, M. Facão, A. Martins, P.S. André, Simulation of fiber fuse effect propagation, in *International Conference on Transparent Networks-Mediterranean Winter 2009*, Angers, France (2009). doi:10.1109/ICTONMW.2009.5385610 (FrP.12)

51. A.M. Rocha, G. Fernandes, F. Domingues, M. Niehus, M. Facão, P.S. André, Halting the fuse discharge propagation using optical fiber microwires. Opt. Expr. **20**(19), 21083–21088 (2012). doi:10.1364/OE.20.021083

52. H.L. Schick, A thermodynamic analysis of the high-temperature vaporization properties of silica. Chem. Rev. **60**(4), 331–362 (1960). doi:10.1021/cr60206a002

53. K. Seo, N. Nishimura, M. Shiino, R. Yuguchi, H. Sasaki, Evaluation of high-power endurance in optical fiber links. Furukawa Rev. **24**, 17–22 (2003)

54. Y. Shuto, Heat conduction modeling of fiber fuse in single-mode optical fibers. IEICE Trans. Commun. B J94-B(8), 928–937 (2011) (in Japanese)

55. Y. Shuto, S. Yanagi, S. Asakawa, M. Kobayashi, R. Nagase, Fiber fuse phenomenon in step-index single-mode optical fibers. IEEE J. Quantum Electron. **40**(8), 1113–1121 (2004). doi:10.1109/JQE.2004.831635

56. Y. Shuto, S. Yanagi, S. Asakawa, M. Kobayashi, R. Nagase, Fiber fuse phenomenon in triangular-profile single-mode optical fibers. J. Lightwave Technol. **24**(2), 846–852 (2006)

57. A. Streek, P. Regenfuß, T. Süß, T, R. Ebert, H. Exner, Laser micro sintering of SiO$_2$ with an NIR-laser, ed. by V.P. Veiko, in *Fundamentals of Laser Assisted Micro- and Nanotechnologies (FLAMN-07), SPIE Proceedings*, vol. 6985, pp. 69850Q (2008). doi:10.1117/12.787121

58. H. Takara, H. Masuda, H. Kanbara, Y. Abe, M. Miyamoto, R. Nagase, T. Morioka, S. Matsuoka, M. Shimizu, K. Hagimoto, Evaluation of fiber fuse characteristics of hole-assisted fiber for high power optical transmission systems, in *Proceedings of the 35th European Conference on Optical, Communication*, p. 312 (2009) (P1.12)

59. K. Takenaga, S. Omori, R. Goto, S. Tanigawa, S. Matsuo, K. Himeno, Evaluation of high-power endurance of bend-insensitive fibers. in *Proceedings of Optical Fiber Communication/National Fiber Optic Engineers Conference* (2008). doi:10.1109/OFC.2008.4528147 (JWA11)

60. K. Takenaga, S. Tanigawa, S. Matsuo, M. Fujimaki, H. Tsuchiya, Fiber fuse phenomenon in hole-assisted fibers, in *Proceedings of the 34th European Conference on Optical Communication*, vol. 5, pp. 27–28 (2008). doi:10.1109/ECOC.2008.4729434 (P.1.14)

61. S. Todoroki, Animation of fiber fuse damage, demonstrating periodic void formation. Opt. Lett. **30**(19), 2551–2553 (2005). doi:10.1364/OL.30.002551

62. S. Todoroki, Origin of periodic void formation during fiber fuse. Opt. Expr. **13**(17), 6381–6389 (2005). doi:10.1364/OPEX.13.006381

63. S. Todoroki, Transient propagation mode of fiber fuse leaving no voids. Opt. Expr. **13**(23), 9248–9256 (2005). doi:10.1364/OPEX.13.009248

64. S. Todoroki, In-situ observation of fiber-fuse ignition, ed. by V.I. Konov, V.Y. Panchenko, K. Sugioka, V.P. Veiko, in *International Conference on Lasers, Applications, and Technologies 2005: Laser-Assisted Micro- and Nanotechnologies, SPIE Proceedings*, vol. 6161, pp. 61610N, 14 May 2005 (2006). doi:10.1117/12.675080

65. S. Todoroki, Threshold power reduction of fiber fuse propagation through a white tight-buffered single-mode optical fiber. IEICE Electron. Expr. **8**(23), 1978–1982 (2011). doi:10.1587/elex.8.1978

66. S. Todoroki, Fiber fuse propagation behavior, in Selected Topics on Optical Fiber Technology, ed. by Y. Moh, S.W. Harun, H. Harun (InTech, Croatia, 2012), pp. 551–570. doi:10.5772/26390 (ISBN 978-953-51-0091-1)

67. S. Todoroki, Partially self-pumped fiber fuse propagation through a white tight-buffered single-mode optical fiber, in *Optical Fiber Communication Conference, OSA Technical Digest*. Optical Society of America (2012). doi:10.1364/OFC.2012.OTh4I.4 (Paper OTh4I.4)
68. S. Todoroki, Fiber fuse propagation modes in typical single-mode fibers, in *Optical Fiber Communication Conference, OSA Technical Digest*. Optical Society of America (2013). doi:10.1364/NFOEC.2013.JW2A.11 (Paper JW2A.11)
69. S. Todoroki, Modes and threshold power of fiber fuse propagation. IEICE Trans. Commun. B J96-B(3), 243–248 (2013) (in Japanese, open access)
70. P.W. Turner, L. Dong, Drawing of silica optical fibers, in *Properties of Glasses and Rare-Earth Doped Glasses for Optical Fibers, EMIS Datareview Series*, ed. by D. Hewak (NSPEC, IEE, London, 1998), pp. 62–64 (ISBN 978-0852969526)
71. A.D. Yablon, *Optical Fiber Fusion Splicing, Springer Series in Optical Sciences* (Springer, Berlin, 2005). doi:10.1007/b137759 (ISBN 978-3-540-23104-2)
72. S. Yanagi, S. Asakawa, M. Kobayashi, Y. Shuto, R. Naruse, Fiber fuse terminator, in *The 5th Pacific Rim Conference on Lasers and Electro-Optics*, vol. 1, Taipei, Taiwan, p. 386. 22–26 July 2003 (2003). doi:10.1109/CLEOPR.2003.1274838 (W4J-(8)-6)

Chapter 2
Fiber Fuse Propagation Modes

Nothing is more revealing than movement.

—Martha Graham

2.1 Introduction

In spite of the 25-year history of fiber fuse research, most researchers still see a fiber fuse as a moving point without an internal structure as summarized in the previous chapter. This is partly because the time for observing this moving object within the scope of their measuring devices is very short. To deepen our knowledge of the fiber fuse, more precise and extensive measurements are needed. In the following three chapters, I try to clarify the inner structure and action of a fiber fuse generated in standard single-mode fibers by focusing on its motion (this chapter), periodic void formation (Chap. 3) and its pump power modulation and response (Chap. 4).

Although the fusing speed is known to increase monotonously with pump power as summarized in Sect. 1.3, the generated damage exhibits various patterns as shown in Fig. 1.8. Thus, the pump power dependence of the following three characteristics is precisely investigated; an in situ image of fiber fuse propagation, void shape, and propagation speed. As a result, it was found that the fusing speed near the propagation threshold deviates from an extrapolated line based on the higher pump power side due to the shrinkage of the confined plasma. This chapter describes the details of the fiber fuse propagation behavior and classifies it into three modes; unstable, unimodal, and cylindrical.

S. Todoroki, *Fiber Fuse*, NIMS Monographs, DOI: 10.1007/978-4-431-54577-4_2,
© National Institute for Materials Science, Japan. Published by Springer Japan 2014

Fig. 2.1 In situ images of fiber fuse propagation captured with an ultrahigh-speed video camera. Original *grayscale* images are converted to *color-scale* images. See Box 2.1 (b) for details

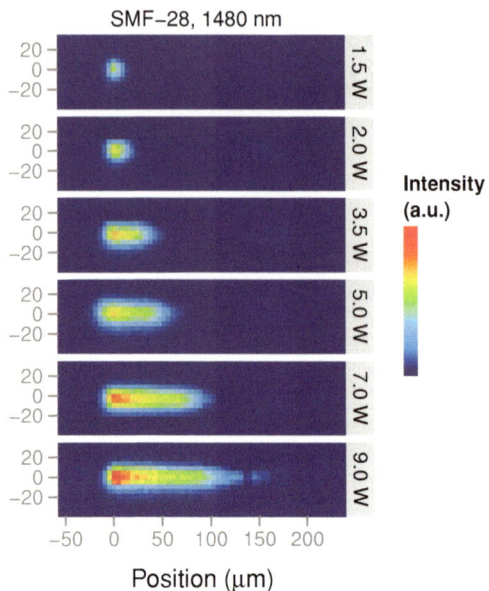

2.2 In Situ Observation of Fiber Fuse Propagation

An orthodox approach to investigate a moving object involves taking a high-speed photograph. Since a fiber fuse moves at roughly 1 m/s, the time for capturing it within the scope of the camera, for a distance of 1 mm for example, is less than 1 ms. Moreover, the exposure time should be as small as possible to avoid overexposure to the strong light emission from the plasma. After preliminary trials in 2004 [1, 2, 4] (see also Box 2.1 (a)), the pump power dependence of the light emission profile was revealed as shown in Fig. 2.1 (see also Box 2.1 (b)), which is a reproduction based on the original data [5].

The fibers used are a commercially produced standard single-mode fiber, SMF-28e (Corning). The light source was a Raman fiber laser (CW, wavelength: 1.48 μm). A fiber fuse was initiated by the local heating of a fiber delivering several watts of light (see Fig. 2.2 (1)). Then, the pump power was reduced to a certain value (2), and the laser oscillation was stopped after the picture taking was completed (3). All the laser operations were performed via a personal computer.

Box 2.1 Videos providing microscopic view of fiber fuse propagation and void formation.

(a) First trial in situ observation in August 2004.

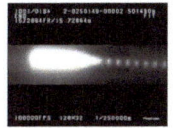   http://imeji.nims.go.jp/collection/8/item/9
Duration: 0:03
Fiber: SMF-28 (Corning)
Pump laser: 1480 nm, 9W
Camera: FASTCAM-MAX (Photron)
Speed: 100,000 fps   Exposure time: 1 μs
References: Fig. 4 in [4]

(b) Pump power dependence of fiber fuse propagation behavior. Original gray-scale images are converted to color-scale images.

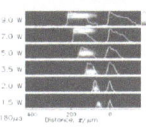   http://imeji.nims.go.jp/collection/8/item/11
Duration: 0:18
Fiber: SMF-28 (Corning)
Pump laser: 1480 nm
Camera: FASTCAM-APX RS (Photron)
Speed: 250,000 fps   Exposure time: 1 μs
References: Fig. 2.1, Fig. 3 in [5]

(c) Fiber fuse propagation and void formation.

   http://imeji.nims.go.jp/collection/8/item/10
Duration: 0:09
Fiber: SMF-28e (Corning)
               immersed in index matching oil
Pump laser: 1480 nm
Camera: FASTCAM SA5 (Photron)
Speed: 30,000 fps   Exposure time: 33.3 μs
References: Fig. 4.8

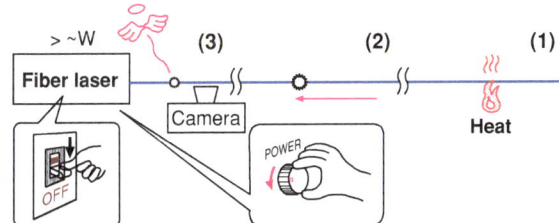

Fig. 2.2 Experimental setup for taking pictures of fiber fuse propagation

Figure 2.3 shows the configuration for taking fiber fuse pictures. The acrylate coating was removed from a segment including the field of vision. The confined plasma propagated from right to left. This is true for all the figures in this book. Since the fiber acts as a cylindrical lens, we should note that the image near the core region is non-linearly expanded in the vertical direction.

Fig. 2.3 Geometrical arrangement of a test fiber and its photograph

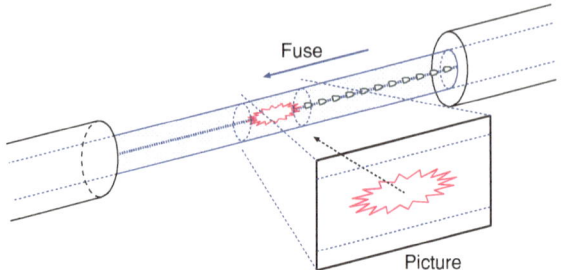

The most important finding from Fig. 2.1 is that there are two types of emission profile along the fiber axis; a symmetric unimodal profile, pumped at 2 W or less, and an asymmetric profile. Their propagation speeds were calculated from successive video images and showed a roughly linear dependence on pump power [5].

2.3 Void Track Variation with Pump Power

Given the difficulty involved in investigating a moving object, the next best subject for investigation is the damage sites that it leaves. The first void of the damage train is the place where confined plasma disappeared due to pump exhaustion. The first void shape variation with pump power was reported in 2005 [5], and Fig. 2.4 reproduces that result with better image quality and enhanced pump power precision. The value before the shutdown was determined with a precision of 0.01 W on the basis of the power stabilization behavior of the light source observed in preliminary experiments. The damage sites in the core region were observed with a digital optical microscope where the samples were immersed in index matching oil.

The volume of the first void increases with the pump power, and its shape changes at about 2 W from a spheroid to a long partially cylindrical void. This tendency agrees well with the in situ observation result shown in Fig. 2.1. These void shapes are expected to retain the outline of the confined plasma if the quenching rate of the glass melt is not delayed by some external conditions including the pump power decay rate. Thus, the decay time of the plasma light emission after the pump laser shutdown was estimated on the basis of ultrahigh-speed photography (see Fig. 2.5). It is no more than $7\,\mu s$ [3], which is far shorter than the time needed for heat diffusion from the core to the cladding surface, that is, a few milliseconds (see Fig. 1.3). Thus, it is reasonable to assume that the plasma outline is frozen into the void shape.

Then, it is possible to classify these plasma shapes into the following three types on the basis of Fig. 2.4 and the void diameter plotted in Fig. 2.6b, c.

1. a spheroid whose diameter increases with the pump power (see Fig. 2.4a and b; × in Fig. 2.6)
2. a spheroid with a constant diameter ((c) and (d); ○)
3. a long partially cylindrical void with a constant diameter ((e)–(h); □).

SMF−28e, 1480 nm

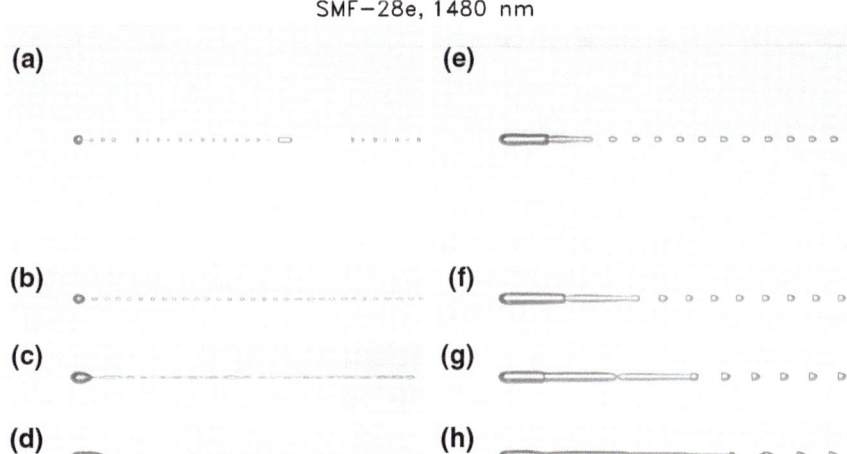

Fig. 2.4 Optical micrographs showing the front part of the fiber fuse damage generated in single-mode silica glass fibers (SMF-28e). The pump laser powers are **a** 1.33 W (propagation threshold), **b** 1.36 W, **c** 1.47 W, **d** 2.03 W, **e** 3.42 W, **f** 5.00 W, **g** 6.99 W, and **h** 8.99 W; wavelength: 1,480 nm. The two thin lines at the *top* and *bottom* of **a** and **e** are the edges of the fiber, whose diameter is 125 mm

Fig. 2.5 In situ images capturing the moment of fiber fuse termination caused by a manual shutdown of the pump laser at $t = 0$ ms. The exposure time was 3.9 µs. The emission had disappeared completely by the second frame after the shutdown. The *donut shape* of the image is because it is of out of focus

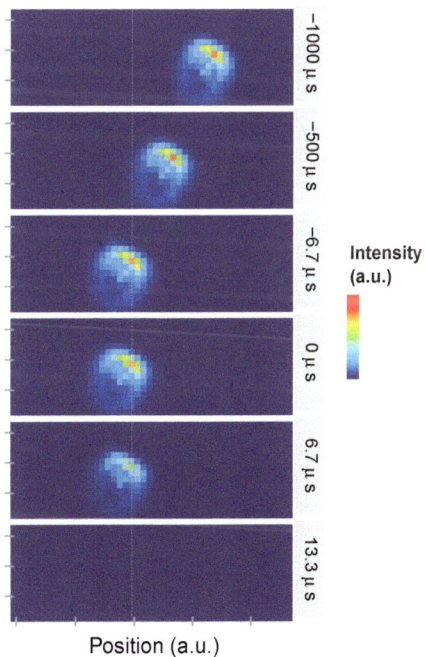

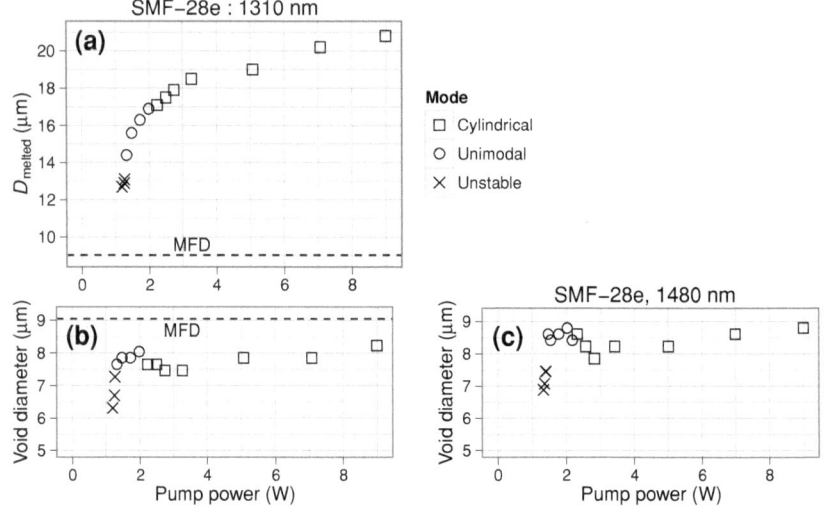

Fig. 2.6 **a** Pump power dependence of the diameter of the modified area in the refractive index and **b** and **c** the first void diameter that remain after the passage of a fiber fuse in SMF-28e. The wavelength is 1,310 nm for (**a**) and (**b**), and 1,480 nm for (**c**)

For explanatory convenience, they are referred to as unstable, unimodal, and cylindrical modes, respectively.

This void diameter saturation in unimodal and cylindrical modes is due to the confined distribution of the pump power at the fiber core. The ceiling value is less than the MFD as shown in Fig. 2.6b. Thus, the plasma length along the fiber increases with the pump power as shown in Figs. 2.1 and 2.4.

At the same time, the plasma temperature is expected to increase. This trend is extracted from the damaged fibers as a cylindrical area surrounding the damage whose refractive index is modified due to the propagation of hot plasma. The outline of this area is seen clearly under tilted illumination as shown in Fig. 1.8. The diameter of this area, D_{melted}, increases monotonously with the pump power as shown in Fig. 2.6a.

The photographs in Fig. 2.4 also show a variation in the void train left after the first void. Periodicity appears in the cylindrical mode (e–h) and in a limited pump power range in the unstable mode (b), whereas it is broken in the unimodal and unstable modes (a, c, d). This will be discussed in detail in the next chapter.

2.4 Precise Measurement of Propagation Speed

The above three-way classification of plasma shape raises the question of whether the fuse propagation speed really increases linearly with the pump power. In fact, a recent high-precision velocity measurement revealed a linearity deviation on the

Fig. 2.7 Pump power dependence of fiber fuse propagation speed in two types of single-mode fibers

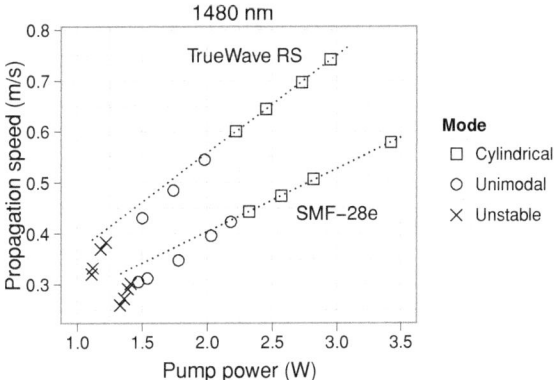

Fig. 2.8 Experimental setup for propagation speed measurement. *PLC* programmable logic controller

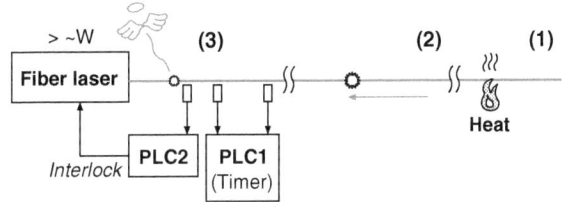

low-energy side as shown in Fig. 2.7 [6, 7]. The propagation speed was calculated from the time difference between two photosensors placed along a fiber 80 mm apart (see Fig. 2.8 (3)). The time unit is 0.1 ms, and the error in the calculated velocities is estimated to be about ±0.01 m/s.

Figure 2.7 shows the results for two types of fibers, SMF-28e and TrueWave RS (OFS), with different MFDs, 10.4 ± 0.5 and 8.4 ± 0.6 (product specifications at 1,550 nm), respectively. The velocity of TrueWave RS is larger than that of SMF-28e owing to its smaller MFD. However, both show similar pump power dependence. In addition, their propagation threshold power values varies in proportion to MFD [7] (see also Sect. 1.4).

Each dotted line is the result of a least squares fit to the four open squares on the right and shows that the remaining points deviate from the line. The deviation becomes larger near the propagation threshold. The slope becomes about three times larger than that shown as a dotted line when the pump power is less than 1.4 W. This behavior is coupled with the void diameter saturation described in the previous section. The cross section of confined plasma perpendicular to the pump beam stops increasing with the pump power at a certain threshold value near 1.4 W. Another slight slope change found at around 2.2 W is due to the geometrical change of the confined plasma from a spheroid to a long partially cylindrical void.

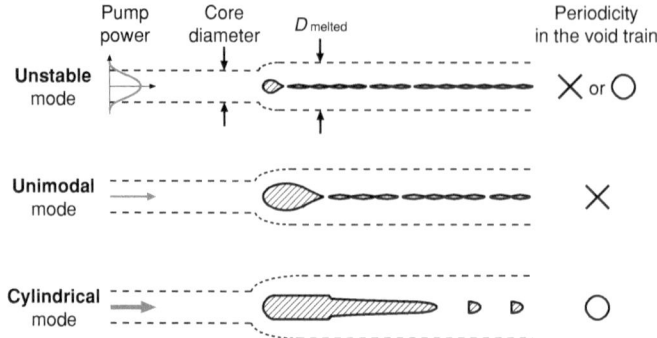

Fig. 2.9 Fiber fuse propagation modes based on the shape and volume of the confined plasma

2.5 Summary

Three types of propagation modes are revealed on the basis of in situ observation and speed measurement of fiber fuse propagation and the damage morphology of fused fibers. The difference comes from the plasma volume confined in the fiber core as summarized in Fig. 2.9.

References

1. I.A. Bufetov, E.M. Dianov, Optical discharge in optical fibers. Phys. Usp. **48**(1), 91–94 (2005). doi:10.1070/PU2005v048n01ABEH002081
2. S. Todoroki, In-situ Observation of Fiber-Fuse Propagation, in *Proceedings of the 30th European Conference Optical Communication Post-deadline papers*, pp. 32–33. Kista Photonics Research Center, Stockholm, Sweden, 2004 (Th 4.3.3)
3. S. Todoroki, Animation of fiber fuse damage, demonstrating periodic void formation. Opt. Lett. **30**(19), 2551–2553 (2005). doi:10.1364/OL.30.002551
4. S. Todoroki, In-situ observation of fiber-fuse propagation. Jpn. J. Appl. Phys. **44**(6A), 4022–4024 (2005). doi:10.1143/JJAP.44.4022
5. S. Todoroki, Origin of periodic void formation during fiber fuse. Opt. Express **13**(17), 6381–6389 (2005). doi:10.1364/OPEX.13.006381
6. S. Todoroki, Fiber Fuse Propagation Modes in Typical Single-Mode Fibers, in *Optical Fiber Communication Conference, OSA Technical Digest* (Optical Society of America, USA, 2013). doi:10.1364/NFOEC.2013.JW2A.11. Paper JW2A.11
7. S. Todoroki, Modes and threshold power of fiber fuse propagation. IEICE Trans. Commun. B **J96–B**(3), 243–248 (2013) (in Japanese)

Chapter 3
Periodic Void Formation

All motion is cyclic. It circulates to the limits of its possibilities
and then returns to its starting point. —'Riches within your
reach!',

Robert Collier

3.1 Introduction

Spatial periodic voids left after the running plasma must have resulted from a temporal periodic action inside the fiber fuse. The interval between the periodic voids, Λ, left in SMF-28e is more than 10 μm and increases with the pump power as shown in Fig. 3.1a. Therefore, from the length of this interval and the corresponding propagation speed, the time required for the formation of one void can be calculated numerically as about 20 μs (see Fig. 3.1b). Although this cycle is long enough to allow us take continuous pictures with state-of-art cameras, the action inside the fiber fuse is still concealed behind the strong light emission (see video clips in Box 2.1).

Alternatively, photographs of fused damage give us some hints of the temporal variation of the internal structure if the effect of structural modification during quenching is carefully excluded. This chapter discusses the mechanism of the bullet-shaped periodic void formation in the cylindrical mode and the small-scale versions that appear in the unstable mode.

3.2 Void Formation Process in Cylindrical Mode

The shape of the first void left by a fiber fuse in cylindrical mode is not always the same [3, 4] as shown in Fig. 3.2. This is because a fiber fuse leaves a void every few tens of microseconds, and the moment that the power is shutdown differs for each time within this cycle.

S. Todoroki, *Fiber Fuse*, NIMS Monographs, DOI: 10.1007/978-4-431-54577-4_3,
© National Institute for Materials Science, Japan. Published by Springer Japan 2014

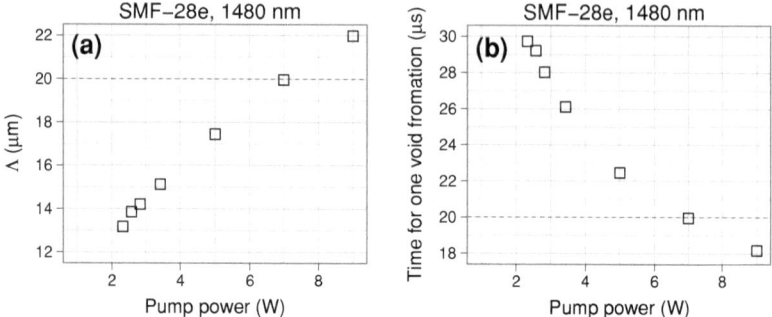

Fig. 3.1 **a** Pump power dependence of periodic void interval, Λ, and **b** calculated time of one void formation left in a SMF-28e fiber pumped at 1,480 nm

Fig. 3.2 Variation of the damage site left by a fiber fuse pumped with ~8.9 W, 1,480 nm light. Vertical lines are drawn at equal intervals of $\Lambda = 21.8\ \mu m$

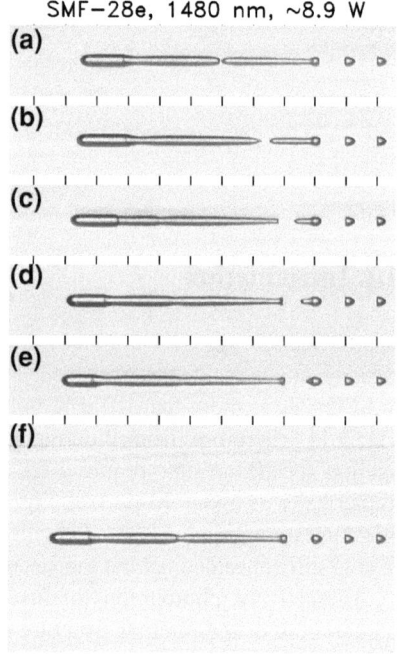

Figure 3.2 shows a collection of fused damage that was frozen from a fiber fuse after a quick shutdown of the pump laser and whose periodic void interval is 21.8 µm. The fiber fuse ran through an acrylate-coated SMF-28e fiber, pumped with ~ 8.9 W, 1,480 nm light. Many samples were prepared and their photographs were sorted in order of increasing distance between the first void and the first periodic void. This sorting operation corresponds to a rearrangement in chronological order within the void formation cycle.

Fig. 3.3 In situ images of fiber fuse propagation through a SMF-28e fiber pumped with 1,480 nm, 9 W light. Speed: 700,000 fps, Exposure time: 0.37 μs. Original grayscale image is converted to a color-scale image

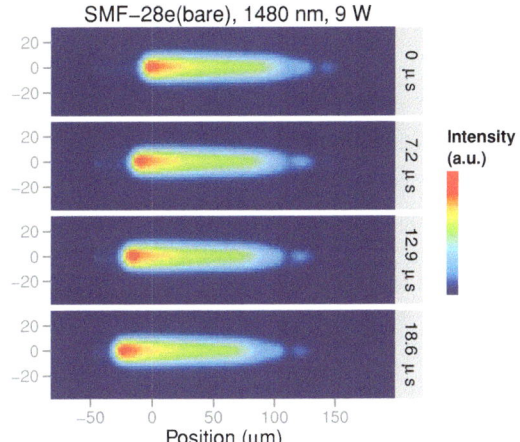

This sequence seemingly shows that the bridge between the first two voids moves to the right while the first void moves toward the light source. Finally, the second void shrinks into a bullet, namely it becomes one of the periodic voids. However, what we have to imagine is the periodic action before quenching. The question is whether or not each frozen bridge appears before quenching.

Since the temperature of a confined plasma is estimated to be several thousand K [1, 2], the surrounding glass melt is subjected to a fairly high pressure. However, the bridges (a) and (b) shown in Fig. 3.2 seem to be fragile as regards such a pressure and must have been formed during quenching in response to the sudden pressure reduction. Namely, a neck appears in the middle of the first void and forms a bridge to compensate for this pressure reduction.[1] In addition, the absence of these two bridges is consistent with an in situ video observation result. Namely, if the bridge appears in the middle of the long plasma before quenching, it interrupts the pump beam to the second void that deactivates the plasma in it. Then, the resulting light emission along the fiber axis will be periodically shortened. However, the width remains almost constant during the cycle of about 18 μs as shown in Fig. 3.3.

Consequently, Fig. 3.2a, b are excluded from the sequence for demonstrating the periodic void formation process. An animation is compiled from the rest of the photographs as shown in Box 3.1a, which clearly reveals how the bullet-like voids are formed. Once a glass bridge appears in the tail, it is pushed backward by the pressure of the confined plasma. At the same time, its viscosity increases because its temperature decreases as the distance from the fusing front increases. In terms of the second void, the rear surface is the first to be frozen and then the front is pushed and becomes flat (see c–f in Fig. 3.2). In other words, this time lag of solidification under the temperature gradient causes the asymmetric void shape.

---

[1] As a matter of fact, this bridge formation during quenching was carefully excluded from the previous discussion on Fig. 2.4e–h assuming that the plasma outline is frozen into a void shape.

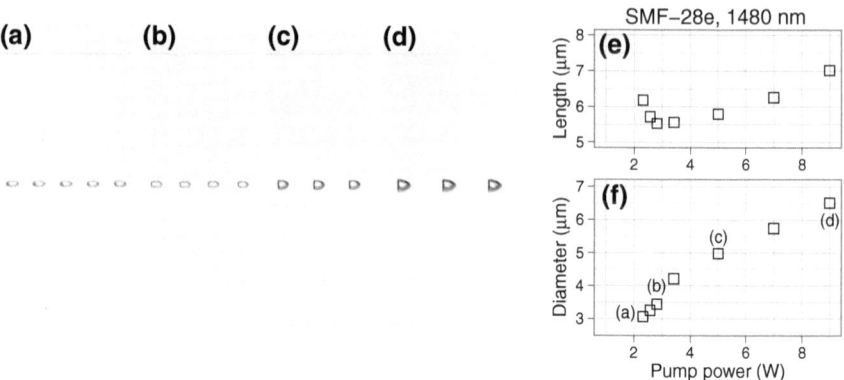

Fig. 3.4 (*Left*) Periodic voids left by a fiber fuse pumped at (**a**) 2.32 W, (**b**) 2.82 W, (**c**) 5.00 W and (**d**) 8.99 W laser (λ: 1,480 nm, fiber: SMF-28e) and (*right*) pump power dependence of their void length (**e**) and diameter (**f**)

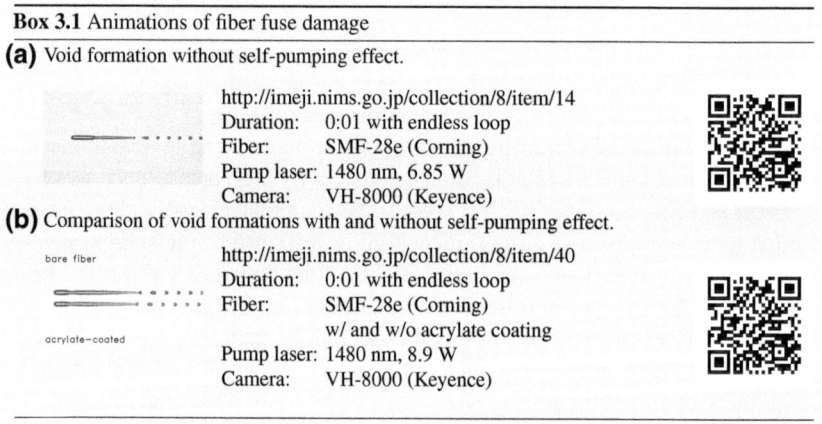

This asymmetry increases with the plasma pressure or pump power as shown in Fig. 3.4a–d. Their diameter and length are plotted in Fig. 3.4e, f. The void diameter expansion is simply related to the pump power increase (see Fig. 3.4f). However, the fall and rise of the void length is a superposition of increasing axial compression and pump power (see Fig. 3.4e).

On the other hand, this asymmetry disappears in unimodal mode. As the pump power decreases, each void in the train is connected each other to lose its discrete periodicity as shown in Fig. 2.4d. At the same time, the plasma loses its tail. Thus, the tail of the confined plasma is an essential factor for discrete periodic void formation.

3.3   Cladding Mode Self-pumping of Confined Plasma

Previous discussions of plasma behavior can be deepened when we consider an additional pumping scheme for a fiber fuse, or cladding mode pumping. Optical fibers are usually coated with transparent acrylate polymer with a refractive index higher than that of cladding silica glass. Without this layer, light that leaks from the core is confined in the cladding layer. This is also true for the fiber fuse emission.

The critical angle of total reflection, θ, is given by Snell's law,

$$\frac{\sin \theta}{\sin 90°} = \frac{n_{\text{air}}}{n_{\text{silica glass}}} = \frac{1}{1.46}, \tag{3.1}$$

at $44°$, i.e., nearly half of a right angle (see Fig. 3.5). Therefore, part of the fiber fuse emission propagates inside the cladding if the propagation angle against the fiber axis is less than $90° - \theta$. Thus, the shortest distance between the starting point of the emission and the intersection of the reflected light and the fiber axis is $2r \tan \theta \simeq 2r$, where r is the fiber radius.

Then, what occurs if the plasma length exceeds the fiber diameter, $2r$? The emission from the top is reflected totally at the cladding surface and reaches the tail of the fiber fuse. Then, the reflected light is absorbed by the plasma and/or surrounding melt containing SiO (see Eq. 1.1). This is a very interesting situation whereby the tail of the fuse can be selectively pumped by the energy dissipated from the fuse itself. Thus, another void formation sequence is compiled based on the damage photographs with $\Lambda = 21.8\,\mu\text{m}$ obtained from a fuse terminated in a bare fiber as shown on the left in Fig. 3.6.

As compared with the acrylate-coated case shown on the right in Fig. 3.6, a slight neck appears at the distance, $2r$, from the void top as indicated with a vertical arrow. In addition, the first void seems to be slightly elongated. This is clearly seen in the animation shown in Box 3.1b.

The first void lengths, l_1, are precisely compared based on the normalized second void length, l_2/Λ (see Fig. 3.7). If the bridge between the first and second voids is assumed to appear during quenching, these two voids are regarded as the first void as illustrated at the top of Fig. 3.7. The value l_2/Λ is regarded as the "phase" of the periodic void formation cycle.

Figure 3.8a shows the relation between these two parameters. The first void length left in a bare fiber ($\bullet$) is 3–4 % longer than that in a buffered fiber ($\circ$). As a matter of fact, this comparison underestimates this elongation effect because the self-pumping effect also makes Λ slightly longer. Its increment is estimated to be 0.2–0.3 μm in a separate experiment described in Sect. 4.2 (see Fig. 4.1). This causes a further elongation of l_1 as shown in Fig. 3.8b. Consequently, the total elongation effect is estimated to be 5–6 %.

This behavior tells us that the confined plasma is elongated owing to the self-pumping of the plasma tail. Namely, the local self-pumping reduces the viscosity of the surrounding melt to allow the plasma tail to extend.

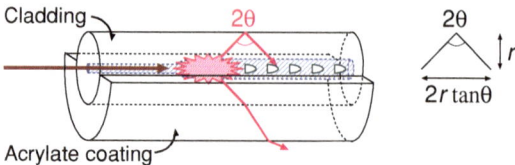

Fig. 3.5 Geometry of cladding mode propagation from a fiber fuse (*top*) and its suppression by acrylate coating (*bottom*)

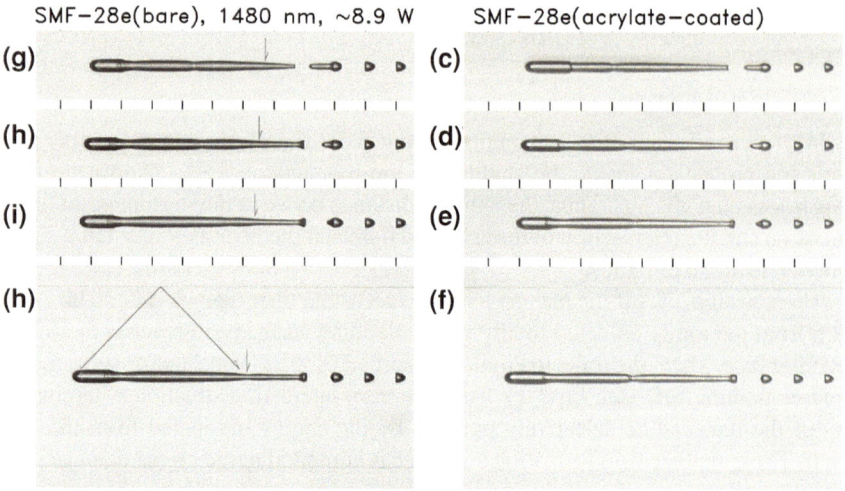

Fig. 3.6 Variation in the damage sites left by a fiber fuse in a bare fiber (*left*) and in an acrylate-coated fiber (*right*, taken from Fig. 3.2). The pump light is 8.9 W, 1.48 μm. The vertical lines are drawn at equal intervals of $\Lambda = 21.8$ μm

Fig. 3.7 Definition of void
train size parameters

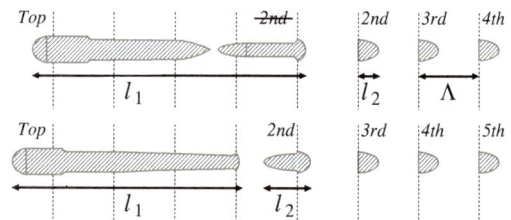

This fact should be taken into account when we examine in situ observation results for fusing through a bare fiber segment as illustrated in Fig. 2.3. The image is not exactly the same as that of a fiber fuse running through an acrylate-coated fiber if the plasma length exceeds the fiber diameter. We have already seen three photographs in this book taken under these circumstances, Fig. 1.5, the bottom of Figs. 2.1 and 3.3, but the effect is insufficiently large to allow us to correct the previous discussions.

Fig. 3.8 Comparison of first void lengths, l_1, for three SMF-28e fibers with different coatings (**a**) and bare fibers with different Λ values (**b**). They are plotted with the normalized length of the second void, l_2/Λ. The pump laser is about 8.9 W, 1,480 nm

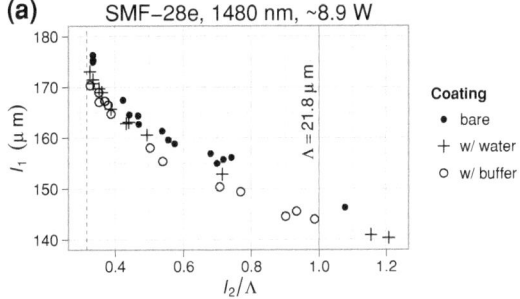

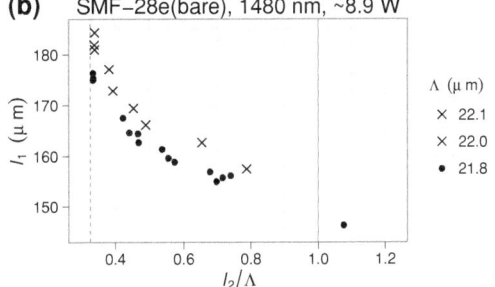

Fig. 3.9 Comparison of the first void lengths, l_1, for two SMF-28e fibers with/without acrylate coating. They are plotted with the normalized length of the second void, l_2/Λ. The pump laser is 6.85 W, 1,480 nm

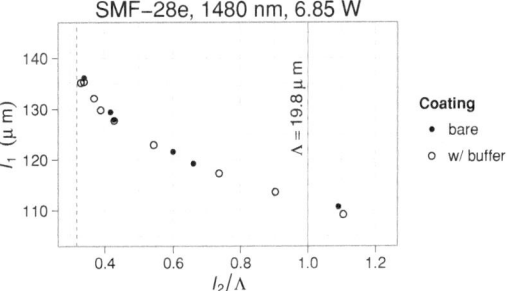

In addition, it is possible to adjust the plasma length if we replace the material surrounding the cladding with one whose refractive index is between those of air and silica glass. One example is a bare fiber immersed in water ($n_{water} = 1.33$, $\theta = 49°$) whose data are plotted with + in Fig. 3.8a. These points are located between the closed and open circles.

However, this self-pumping effect disappears when the plasma length is shorter than $2r \tan \theta$. Figure 3.9 demonstrates this situation. In contrast to Fig. 3.8a, the fiber coating does not affect the first void length when the pump power is 6.85 W. Thus, the critical power is expected to be between 6.85 and 8.9 W. This value becomes smaller when we use bare fibers with a diameter less than 125 μm and tapered fibers without a coating.

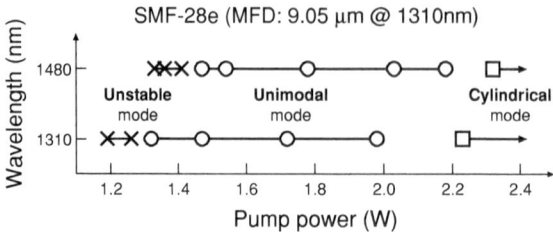

Fig. 3.10 Experimental condition map for three propagation modes for SMF-28e fiber based on pump power and wavelength

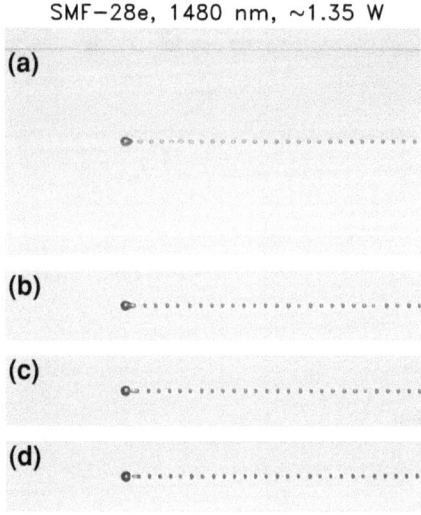

Fig. 3.11 Differences in the damage sites left by a fiber fuse in the unstable mode pumped with about 1.3 W, 1,480 nm light

3.4 Periodic Void Formation in Unstable Mode

A fiber fuse in the unstable mode appears with a very narrow range of pump power, for example, from 1.19 W (propagation threshold for SMF-28e fiber with 9.05 μm in MFD at 1,310 nm) [6, 7] to 1.26 W (or more, but less than 1.32 W, see also Fig. 3.10). However, it leaves different damage patterns depending on the pump power. They are classified into the following three types in decreasing order of pump power:

1. a train of thin filaments; similar to Fig. 2.4c
2. periodic voids; see Figs. 1.8b and 2.4b
3. voids with broken periodicity; see Fig. 2.4a.

The mechanism of this periodic void formation must be different from that in the cylindrical mode because the confined plasma in the unstable mode does not have any apparent tail. Thus, a void formation sequence is compiled from photographs left

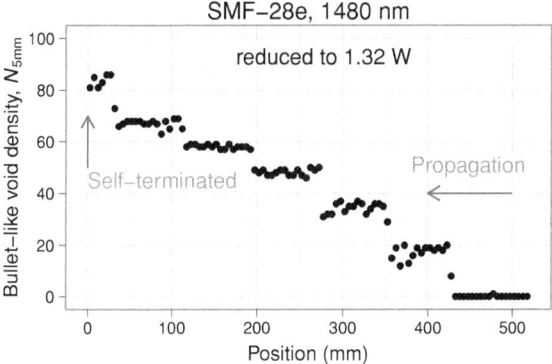

Fig. 3.12 Variation in the number of bullet-like voids along SMF-28e fiber before the self-termination of a fiber fuse. The pump laser power was reduced step wise to 1.32 W

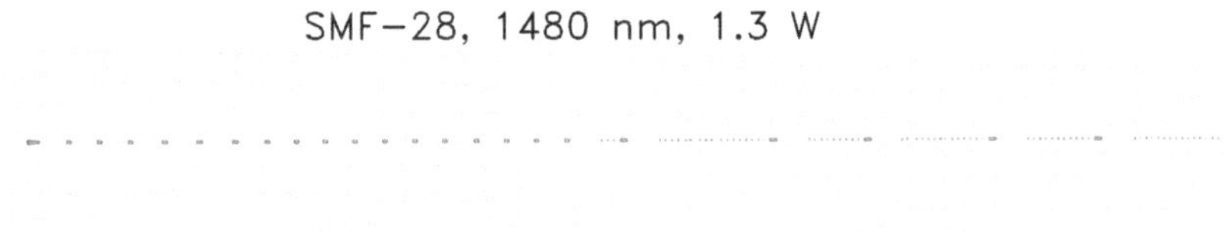

SMF−28, 1480 nm, 1.3 W

Fig. 3.13 Example of void train left by a fiber fuse just before the self-termination. In situ video image of the *right half* is shown in Box 4.1

by a fiber fuse pumped with 1,480 nm light as shown in Fig. 3.11. It clearly shows that a tail appears transiently during the void formation cycle and is pinched off to form one of the periodic voids [5].

This exquisite performance is easily broken by a slight modulation of the pump power. As the power is reduced, the fiber fuse begins to insert an irregular pattern consisting of a short void-free segment and a single bullet-like void into the periodic void train (see Fig. 2.4a). The frequency of this insertion increases as the pump power is decreased until the fiber fuse disappears at the propagation threshold power [6, 7]. Figure 3.12 shows this tendency quantitatively with a parameter N_{5mm}, which is the density of the single bullet-like voids per 5 mm, whose behavior is coupled with the stepwise reduction of the pump power from about 1.35 W ($N_{5mm} = 0$) to the propagation threshold (1.33 W). The mechanism of this irregularity formation is discussed in Sect. 4.3.

At the moment just before self-termination, a fiber fuse often leaves a third type of periodic void as shown in Fig. 3.13. This is the result of the successive appearance of the defect pair.

This pair also appears near a fusion splice point where the waveguide structure is slightly modified. The resulting pump beam modulation brings about a slight reduction in the feed to the fiber fuse. An example is shown in Fig. 3.14. The splice point is located at a hollow sphere in the cladding, which was accidentally captured

SMF—28e, 1480 nm, ~1.35 W

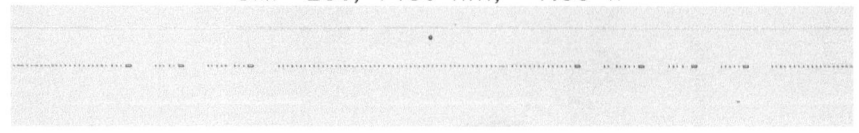

Fig. 3.14 Example of void-free segments that appear around a splice point located at a hollow sphere in the cladding

during splicing. Sometimes a splice point captures a fiber fuse when the pump power is sufficiently near the propagation threshold as shown the video in Box 3.2.

Box 3.2 Fiber fuse self-termination in unstable mode

- A fiber fuse pumped at 1.30 W self-terminates near a fusion splice point in a bare segment. It is slightly activated by the self-pumping effect described in Sec. 4.2 during the propagation through the white tight-buffered segment. However, its energy gain is estimated to be as small as 0.001 W, corresponding to $\Delta N_{5\ mm} = 4$ (see Fig. 2 in [7] and Fig. 3.12).

   http://www.youtube.com/watch?v=6wR7ZsyQHt0
Duration: 0:18
Fiber: White tight-buffered SMF-28 (Corning)
Pump laser: 1480 nm, ~ 1.3W
Camera: Canon VB101
Speed: 30 fps
References: Fig. 2 in [5]

3.5 Summary

The origin of periodic void formation is a tail behind the confined plasma. Its periodic separation is observed in the cylindrical and unstable modes and demonstrated through a sequence of photographs in which the first void is frozen at different moments in the void formation cycle. The strong light absorbance of the fiber fuse is confirmed through cladding mode self-pumping.

References

1. E.M. Dianov, V.E. Fortov, I.A. Bufetov, V.P. Efremov, A.E. Rakitin, M.A. Melkumov, M.I. Kulish, A.A. Frolov, High-speed photography, spectra, and temperature of optical discharge in silica-based fibers. IEEE Photon. Technol. Lett. **18**(6), 752–754 (2006). doi:10.1109/LPT.2006.871110
2. D.P. Hand, J. St, P. Russell, Solitary thermal shock waves and optical damage in optical fibers: the fiber fuse. Opt. Lett. **13**(9), 767–769 (1988). doi:10.1364/OL.13.000767

3. S. Todoroki, Animation of fiber fuse damage, demonstrating periodic void formation. Opt. Lett. **30**(19), 2551–2553 (2005). doi:10.1364/OL.30.002551
4. S. Todoroki, Origin of periodic void formation during fiber fuse. Opt. Expr. **13**(17), 6381–6389 (2005). doi:10.1364/OPEX.13.006381
5. S. Todoroki, Transient propagation mode of fiber fuse leaving no voids. Opt. Expr. **13**(23), 9248–9256 (2005). doi:10.1364/OPEX.13.009248
6. S. Todoroki, Threshold power reduction of fiber fuse propagation through a white tight-buffered single-mode optical fiber. IEICE Electron. Expr. **8**(23), 1978–1982 (2011). doi:10.1587/elex.8.1978
7. S. Todoroki, S, Partially self-pumped fiber fuse propagation through a white tight-buffered single-mode optical fiber, in *Optical Fiber Communication Conference, OSA Technical Digest*. Optical Society of America (2012). doi:10.1364/OFC.2012.OTh4I.4. Paper OTh4I.4

Chapter 4
Delayed Response of Silica Melt to Pump Modulation

On ne voit bien qu'avec le coeur. L'essentiel est invisible pour
les yeux. — 'Le Petit Prince',

Antoine de Saint-Exupéry

It is only with the heart that one can see rightly; what is
essential is invisible to the eye. — 'The Little Prince',

Katherine Woods (trans.)

4.1 Introduction

In the previous two chapters, we revealed the inner structure of a fiber fuse and
its action on the basis of experimental results including speed measurements, in situ
observations, and a statistical analysis of the first void structure. However, the discussions are all only valid for static conditions, i.e., when the pump power is constant or
changes slowly enough to remain in equilibrium with the exterior atmosphere. Then,
what if the pumping scheme changes faster than the relaxation of the surrounding
viscous melt of silica glass?

To answer this question, we have to trace the fiber fuse behavior on a time-series
basis longer than the periodic void formation cycle of about 20 μs. This cycle is
the fundamental time unit that helps us to detect the time-varying behavior of a
fiber fuse because its delayed response is fully encoded as the spatial modification of the void train. Thus, this chapter focuses on decoding the void sequences
that were tentatively modulated by self-pumping, pump laser power, and waveguide
structure. These modulations produce several irregular void patterns, some of which
have already been reported elsewhere without their formation mechanism. However,
the delayed response of silica melt is found to be a clue to understanding these
mechanisms.

S. Todoroki, *Fiber Fuse*, NIMS Monographs, DOI: 10.1007/978-4-431-54577-4_4,   37
© National Institute for Materials Science, Japan. Published by Springer Japan 2014

4.2 Self-pumping Modulation

As discussed in Sect. 3.2, the self-pumping effect appears when a long-tailed fiber fuse enters a bare fiber in air. The selective activation of a plasma tail causes not only plasma elongation but also a slight extension of the periodic void interval, Λ. Figure 4.1 shows the positional dependence of Λ over a bare fiber segment in a acrylate-coated SMF-28e fiber. The Λ values are determined by employing the image processing of digitized damage photographs. Clear small changes are detected at both ends of the bare segment.

This effect can be enhanced by coating the bare segment with white oil paint because nearly all the visible light emission from the fiber fuse is scattered by white pigments on the cladding surface and some of them return to the fiber fuse to be absorbed again [12, 13] (see Fig. 4.2). Moreover, this enhancement is not only limited to a long-tailed fuse but also occurs in all three propagation modes.

Figure 4.3 shows another positional dependence of Λ over a white-painted segment in a partially stripped SMF-28e fiber. In this case, the total-reflection-based self-pumping is suppressed due to the short plasma tail, namely the pump power of 6.9 W is comparable with the similar case described in Sect. 3.3 (see Fig. 3.9). As a result, $\Delta\Lambda$ is roughly three times larger than in the previous case without white oil paint shown in Fig. 4.1. The extended void interval between (2) and (3) is about 20.6 μm. The same value is realized by pumping with 7.6 W without white oil paint as calculated from the data shown in Fig. 3.1a. However, the energy distribution along the fiber axis is expected to be different in these two cases, i.e., with/without self-pumping.

One common feature in these traces is the gradual transition and overshooting/undershooting of Λ after passing one of the borders. Figure 4.3c shows a clear undershooting, which ends about 0.7 mm (Δx) from the transition onset. It takes 0.7 ms for the fiber fuse to propagate. This delay is due to the high viscosity of the silica melt surrounding the plasma. The relaxation time of this plastic flow is considered to be of sub-millisecond order.

This delayed response is indispensable as regards the action of fiber fuse terminators based on the pump power density reduction listed in Table 1.5. A fiber fuse is terminated if the MFD expansion occurs faster than the plastic flow of the surrounding melt, and the resulting energy feed to the plasma is reduced immediately. This condition is expected to be satisfied when the modification length is far shorter than $\Delta x = 0.7$ mm.

4.3 Pump Power Modulation

Compared with the self-pumping modulation, the direct modulation of the pump power induces a greater change in the void formation. When the laser driver controls the output power so that it increases/decreases step-wise as shown in Fig. 4.4,

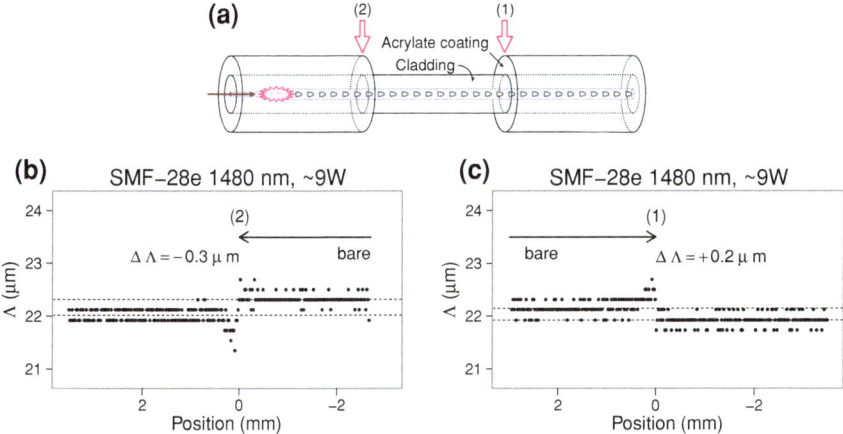

Fig. 4.1 **a** Structure of the test fiber and **b** and **c** positional dependence of void interval, Λ. *Horizontal dashed lines* show the average values before and after the border (*1*) or (*2*). The input laser power is about 9 W

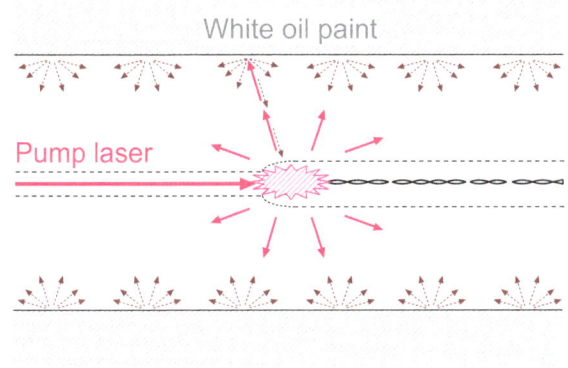

Fig. 4.2 Geometry of backscattering from *white* oil paint at the cladding surface

the plasma leaves an irregular segment at each modulation step [14]. Some typical examples are shown in Fig. 4.5. The decrement/increment of the pump power is completed in 0.2 ms as shown in Fig. 4.4c.

A short void-free segment appears at the moment the pump power increment, ΔP, occurs in all three propagation modes. The segment length increases with ΔP and is unrelated to the estimated distance traveled in 0.2 ms [14]. When we consider that the light–heat conversion is much faster than the plastic flow, this temporal termination of the void formation results from a delayed volume expansion of the hollow glass melt and the preceding temperature increase of the glass melt surface that enables the fiber fuse to completely fill the vacancy behind it before being quenched (see left

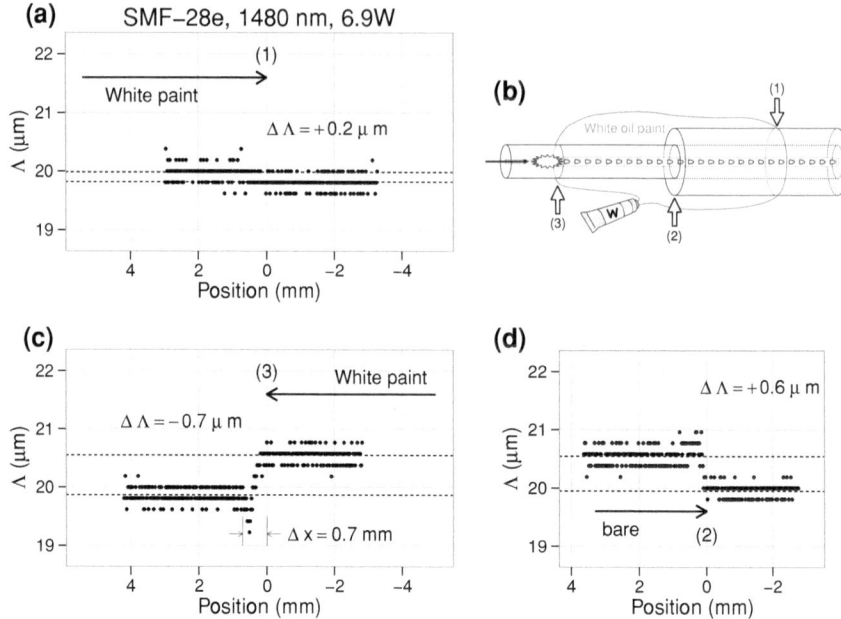

Fig. 4.3 **b** Structure of the test fiber and **a**, **c**, and **d** positional dependence of the void interval, Λ. *Horizontal dashed lines* are the average values before and after the border (*1*), (*2*), or (*3*). The input laser power is about 6.9 W

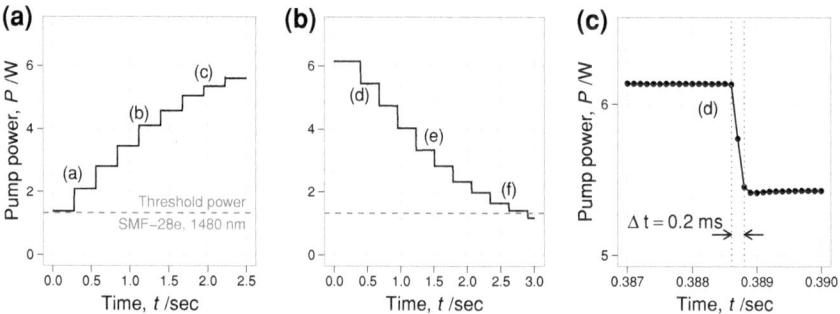

Fig. 4.4 Examples of step-wise *increments/decrements* of the pump power needed to make the irregular void sequences shown in Fig. 4.5a–f

side of Fig. 4.6). The void formation restarts when the delayed volume expansion is completed to reach its equilibrium state.

On the other hand, the power decrement causes the insertion of a long irregular void. This is because the immediate temperature fall does not allow a synchronized volume reduction in the hollow melt and freezes its rear part to form an irregular void (see the right side of Fig. 4.6).

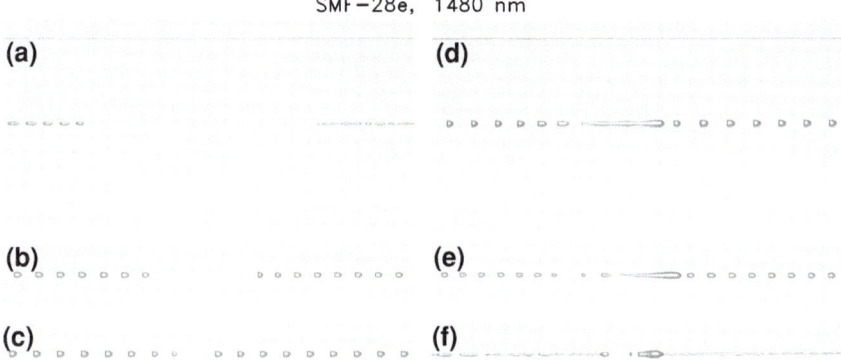

Fig. 4.5 Photographs of irregular segments left at the step-wise power modulation points (a)–(f) shown in Fig. 4.4

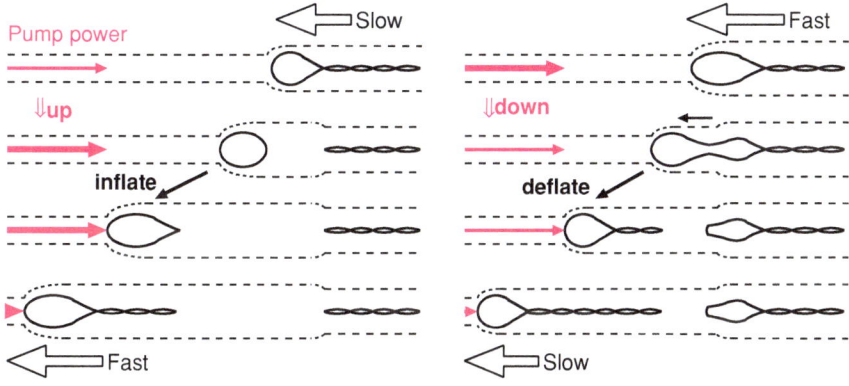

Fig. 4.6 Formation mechanisms of irregular void patterns shown in Fig. 4.5

There have been reports describing a long irregular void inserted in periodic voids that appear as in Fig. 4.5d [1, 5–7]. Considering that the pump laser in this research was either an Ar or a Nd:YAG laser, those irregular voids were possibly generated by an unintentional pump power modulation caused by dust floating in the light path or the absence of a power stabilizer.

There is another suggestion regarding this long irregular void [1], in which the periodic necking in the void constitutes the capture of the "incipient process" of periodic void formation driven by Rayleigh instability. Although this effect well explains the formation mechanism of the water droplets from a tap and bubbles through a water flow as illustrated in Fig. 4.7, this idea seems unlikely for the present case when we notice the high viscosity of surrounding silica melt [15] located along a steep temperature gradient.

Fig. 4.7 Examples of jet breakdown caused by Rayleigh instability, **a** a jet of water broken up into droplets (see [8] p. 131) and **b** a bubble train in a water flow (see [4] p. 540)

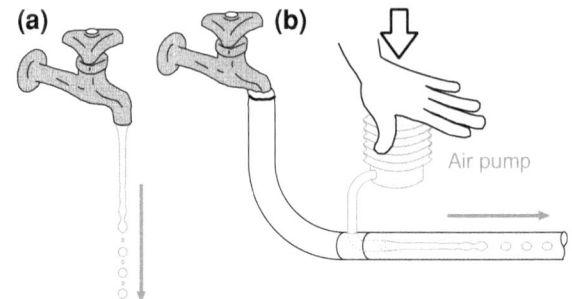

A void-free segment also appears with a bullet-like void in the unstable mode as discussed in Sect. 3.4. This defect pair also resulted from the nature of the silica melt. The evidence is a glint of light that appears when a fiber fuse generates this defect pair (see Box 4.1) [9].

Box 4.1 Fiber fuse propagation in unstable mode (in situ)

- Recorded just before self-termination (see Fig. 3.13). Original gray-scale images are converted to color-scale images.

http://imeji.nims.go.jp/collection/8/item/13
Duration: 0:54
Fiber: SMF-28 (Corning)
Pump laser: 1480 nm, 1.3W
Camera: FASTCAM-APX RS (Photron)
Speed: 250,000 fps   Exposure time: 1 μs
References: Figs. 7, 8 and 9 in [9]

Figure 4.8 shows a time series of in situ photographs in which both light emission and voids are recorded. This is extracted from the video clip shown in Box 2.1 (c). The vertical arrow indicates the position where the emission is maximized ($t = 0.100$ ms). The photographs clearly show that the maximum point is located at the border between the void-free segment and the bullet-like void. It is reasonable that the void-free segment appears as the emission increases and the bullet is formed as it decreases if we assume that the hollow glass melt inflates and deflates successively. In fact, a tentative heat emission at the void-free segment is recorded as a slight expansion of D_{melted}, or the area of modified refractive index, as shown in Fig.1.8 a.

Figure 4.9 shows another sequence of quenched samples sorted to reproduce the plasma propagation along the defect pair. The inflation and deflation of the confined plasma are frozen in sequences (a)–(c) and (c)–(e), respectively.

On the other hand, one question still remains, namely why does the fiber fuse release a light/heat pulse while the pump power is constant? A possible explanation is that the transient tail formation shown in Fig. 3.11 becomes unstable compared with the plasma expansion as the pump power approaches the propagation threshold.

Fig. 4.8 In situ images of a fiber fuse in the unstable mode through a SMF-28e fiber pumped at 1,480 nm, ∼1.3 W light. Speed: 30,000 fps, exposure time: 33.3 μm. The *vertical arrow* indicates the point of maximum emission at $t = 0.100$ ms

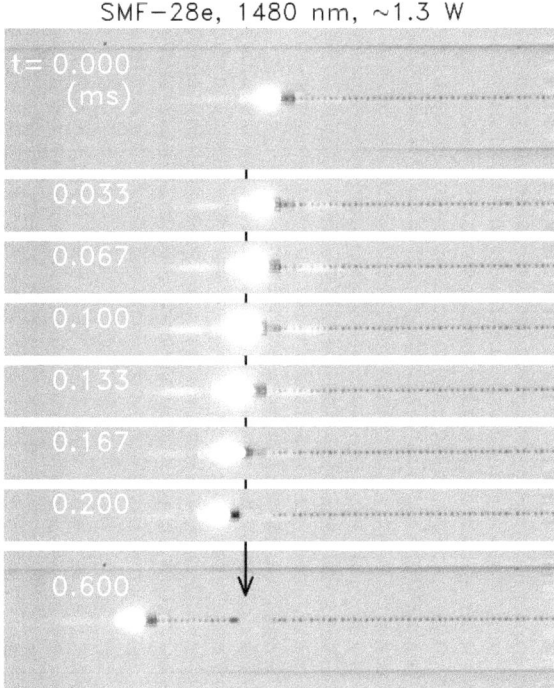

However, the expansion soon stops because the pump power cannot maintain the temperature of surrounding melt. Consequently, the rear part of the melt becomes frozen (see Fig. 4.9d). Then, the plasma detaches a bullet-like void and resumes the formation of periodic voids (see Fig. 4.9e).

Another void-free segment is found just after fiber fuse initiation [9, 10] as shown in Fig. 1.6. It is natural to consider that this formation mechanism is similar to the previous case. The periodic voids are generated only after the hollow melt reaches an equilibrium state with the confined plasma and its energy flow.

4.4 Waveguide Modulation

The irregular segments discussed above, namely a void-free segment and a long void, also appear at a fusion splice point between fibers with different MFDs [11]. Figures 4.10 and 4.11 show some examples with SMF-28e and LEAF fibers (Corning) whose MFDs at 1,550 nm are 10.4 ± 0.5 and 9.6 ± 0.4 μm (product specifications), respectively. Their formation mechanisms can be also deduced in terms of the pump modification and the delayed response.

SMF−28e, 1480 nm, ~1.35 W

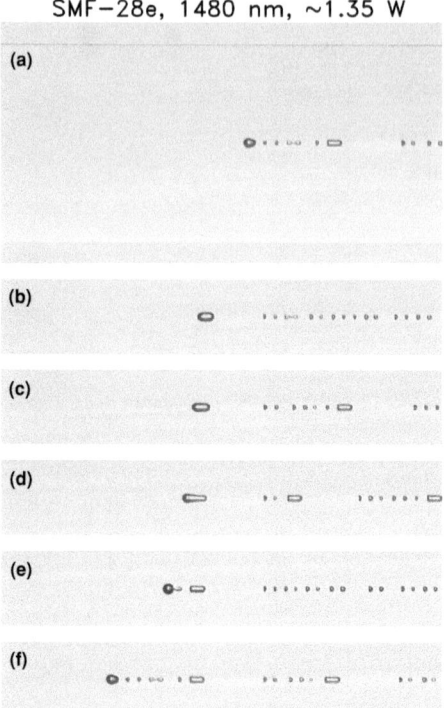

Fig. 4.9 A series of damage sites left by a fiber fuse in the unstable mode pumped at 1.3 W, 1,480 nm light. They are sorted to reproduce the fiber fuse propagation along the void-free segment at the center

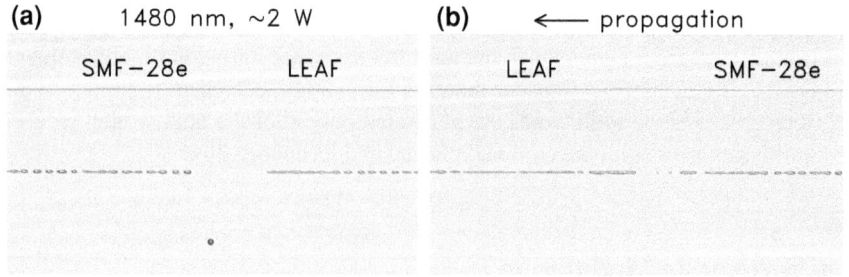

(a)   1480 nm, ~2 W   **(b)**   ⟵ propagation

SMF−28e   LEAF   LEAF   SMF−28e

Fig. 4.10 Photographs of irregular segments left by a fiber fuse in the unimodal mode propagating over fused splice points between SMF-28e and LEAF fibers (Corning). The *splice point* is located at a hollow sphere in the cladding on the *left*, which was accidentally captured during the fusion splice operation

When a fiber fuse passes over a splice from the small to the large MFD side, its waist widens as it loses speed and void formation stops until it fits on the new waveguide as shown on the left in Fig. 4.12. This is shown in Figs. 4.10a and 4.11b.

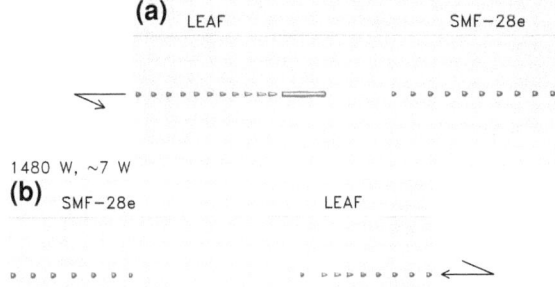

Fig. 4.11 Photographs of irregular segments left by a fiber fuse in the cylindrical mode propagating over fused *splice points* between SMF-28e and LEAF fibers (Corning)

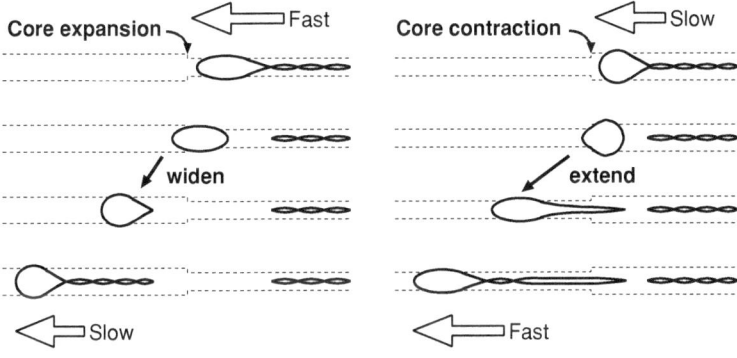

Fig. 4.12 Formation mechanisms of irregular void patterns shown in Fig. 4.10

In the opposite case, the waist of the confined plasma tightens as it gains speed. The resulting long tail is frozen to form a long void as shown on the right in Fig. 4.12 (see also Figs. 4.10b and 4.11a).

Figure 4.13 shows in situ observation results obtained over splice points between SMF-28e and IH 1060 fibers (Corning). Their MFD values are 9.2 ± 0.4 μm @ 1,310 nm and 6.2 ± 0.3 μm @ 1,060 nm (product specifications), respectively. The speed of the plasma clearly decreases/increases as the plasma shortens/lengthens, respectively. Simultaneously, the light emission increases/decreases and the plasma length slightly decreases/increases, respectively (see also Box 4.2). The resulting void trains show similar tendencies as discussed above (see the photographs in the video in Box 4.2).

Fig. 4.13 Intensity profiles of plasma propagating over the *splice points* of SMF-28e and HI1060 fibers (see Box 4.2)

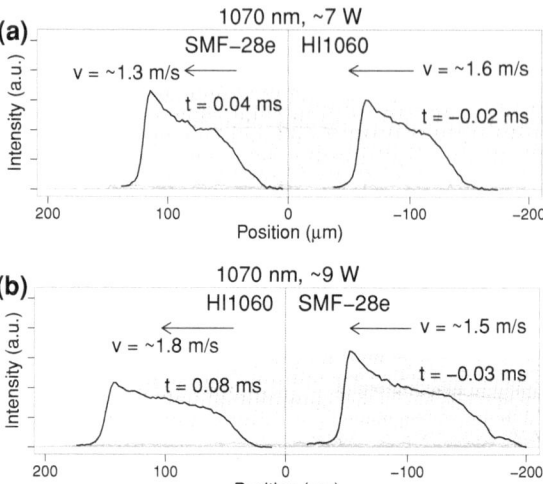

Box 4.2 Fiber fuse running over hetero-core splice point (in situ)

Original gray-scale images are converted to color-scale images.

(a) A fiber fuse propagating over the splice point from HI1060 to SMF-28e fibers where a large spherical void is located in the cladding.

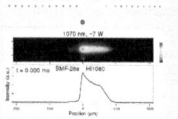   http://imeji.nims.go.jp/collection/8/item/41
Duration: 0:11
Fiber: SMF-28e and IH 1060 (Corning)
Pump laser: 1070 nm, 7W
Camera: FASTCAM SA5 (Photron)
Speed: 525,000 fps   Exposure time: 0.37 μs

(b) A fiber fuse propagating over the splice point from SMF-28e to HI 1060 fibers where a large spherical void is located in the cladding.

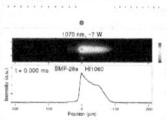   http://imeji.nims.go.jp/collection/8/item/42
Duration: 0:10
Fiber: SMF-28e and IH 1060 (Corning)
Pump laser: 1070 nm, 9W
Camera: FASTCAM SA5 (Photron)
Speed: 325,500 fps   Exposure time: 0.37 μs

These two defect structures appear by turns in a two-mode fiber allowing the LP_{01} and LP_{02} modes of the pump laser to propagate. Bufetov et al. [3] reported a periodic damage sequence consisting a single long void, a void-free segment, and a few bullet-like voids. This sequence is surrounded by a quenched glass layer that is modulated by the interference of the two modes. Namely, d_{melted} varies with the intensity distribution of the pump power along the core. In addition, the void-free segment and the long void appear in the segments where d_{melted} increases/decreases, respectively. This behavior agree with the cases examined in this section.

In addition, another type of irregularity is known to appear near a hetero-core splice [11] and in a few-mode fiber [2]. This irregularity consists of bullet-like voids whose pointing direction is opposite to that if the regular periodic voids as discussed in Chap. 3 (see the photograph in the video in Box 4.2 (b)). Although their formation mechanisms have yet to be clarified, they will be revealed by analyzing a large number of damage photographs in terms of the delayed response of the silica melt.

4.5 Summary

A precise comparison of the time-varying pumping procedure and the resulting void formation revealed an essential characteristic of the fiber fuse in which the confined plasma is surrounded by a viscous glass melt. The delayed response is found to be of sub-millisecond order. Moreover, the delayed inflation/deflation of the melt results in a void-free segment and a long void, respectively, which are included in some previously reported irregular void sequences.

References

1. R.M. Atkins, P.G. Simpkins, A.D. Yablon, Track of a fiber fuse: a rayleigh instability in optical waveguides. Opt. Lett. **28**(12), 974–976 (2003). doi:10.1364/OL.28.000974
2. I.A. Bufetov, A.A. Frolov, A.V. Shubin, M.E. Likhachev, C.V. Lavrishchev, E.M. Dianov, Fiber fuse effect: new results on the fiber damage structure, in *Proceedings of the 33rd European Conference on Optical Communication*, vol. 1, pp. 79–80. IEE's Photonics Professional Network, Berlin, Germany (2007), (Mon 1.5.2)
3. I.A. Bufetov, A.A. Frolov, A.V. Shubin, M.E. Likhachev, S.V. Lavrishchev, E.M. Dianov, Propagation of an optical discharge through optical fibres upon interference of modes. Quantum Electron. **38**(5), 441–444 (2008). doi:10.1070/QE2008v038n05ABEH013751
4. S. Chandrasekhar, in *Hydrodynamic and Hydromagnetic Stability*. International Series of Monographs on Physics (Oxford, England), (Dover Publications, USA, 1981) (ISBN 978-0486640716)
5. D.D. Davis, S.C. Mettler, D.J. DiGiovani, Experimental Data on the Fiber Fuse, in: *SPIE Proceedings 27th Annual Boulder Damage Symposium: Laser-Induced Damage in Optical Materials: 1995*, vol. 2714, ed. by H.E. Bennett, A.H. Guenther, M.R. Kozlowski, B.E. Newnam, M.J. Soileau, pp. 202–210. SPIE (1996), (Boulder, CO, USA, 30 Oct 1995). doi:10.1117/12.240382
6. T.J. Driscoll, J.M. Calo, N.M. Lawandy, Explaining the optical fuse. Opt. Lett. **16**(13), 1046–1048 (1991). doi:10.1364/OL.16.001046
7. D.P. Hand, P.S.J. Russell, Solitary thermal shock waves and optical damage in optical fibers: the fiber fuse. Opt. Lett. 13(9), 767–769 (1988). doi:10.1364/OL.13.000767
8. C. Isenberg, *The Science of Soap Films and Soap Bubbles*, new edn. (Dover Publications, USA, 1992) (ISBN 978-0486269603)
9. S. Todoroki, Transient propagation mode of fiber fuse leaving no voids. Opt. Express **13**(23), 9248–9256 (2005). doi:10.1364/OPEX.13.009248

10. S. Todoroki, In-Situ Observation of Fiber-Fuse Ignition, in *International Conference on Lasers, Applications, and Technologies 2005: Laser-Assisted Micro- and Nanotechnologies, SPIE Proceedings*, vol. 6161, ed. by V.I. Konov, V.Y. Panchenko, K. Sugioka, V.P. Veiko pp. 61,610N-1-4. SPIE (2006), (St. Petersburg, Russia, 14 May 2005, LSK3). doi:10.1117/12.675080

11. S. Todoroki, In situ observation of modulated light emission of fiber fuse synchronized with void train over hetero-core splice point. PLoS One **3**(9), e3276 (2008). doi:10.1371/journal.pone.0003276

12. S. Todoroki, Threshold power reduction of fiber fuse propagation through a white tight-buffered single-mode optical fiber. IEICE Electr. Express **8**(23), 1978–1982 (2011). doi:10.1587/elex.8.1978

13. S. Todoroki, Partially Self-Pumped Fiber Fuse Propagation Through a White Tight-Buffered Single-Mode Optical Fiber, in *Optical Fiber Communication Conference, OSA Technical Digest*. Optical Society of America, USA (2012). doi:10.1364/OFC.2012.OTh4I.4. Paper OTh4I.4

14. S. Todoroki, Fiber Fuse Propagation Modes in Typical Single-Mode Fibers, in *Optical Fiber Communication Conference, OSA Technical Digest*. Optical Society of America (2013). doi:10.1364/NFOEC.2013.JW2A.11. Paper JW2A.11

15. S.I. Yakovlenko, Plasma behind the front of a damage wave and the mechanism of laser-induced production of a chain of caverns in an optical fibre. Quantum Electron. **34**(8), 765–770 (2004). doi:10.1070/QE2004v034n08ABEH002845

Chapter 5
Conclusion

La verdadera ciencia enseña, por encima de todo, a dudar y a ser ignorante. —"Del Sentimiento Trágico de la Vida",

Miguel de Unamuno y Jugo

True science teaches, above all, to doubt and to be ignorant. —"Tragic Sense of Life",

J. E. Crawford Flitch (trans.)

For a quarter of a century since its discovery, the fiber fuse has been recognized as a moving point without an internal structure as briefly summarized in Chap. 1. Through the discussions provided in the subsequent three chapters, a better model is proposed consisting of a hollow silica glass melt confining plasma and is found to be effective in describing fiber fuse propagation and void formation in typical single-mode fibers.

In Chap. 2, the fiber fuse behavior is shown to vary with the shape and volume of the confined plasma with respect to the pump beam size and is classified into three propagation modes.

In Chap. 3, the temporal periodic action of the hollow silica melt is revealed through a statistical analysis over a series of fused damage structures frozen at different moments within the periodic void formation cycle. This methodology enables us to discuss the dynamic behavior of the confined plasma based on a collection of static frozen void structures. As an example, the elongation of the confined plasma by cladding-mode self-pumping is demonstrated.

In Chap. 4, the delayed response of the silica melt to pump power modulation is found to be the origin of some previously reported irregular void structures. This system is unique in the sense that the temporal variation of the energy flow is recorded in a spatial sequence of fused damage along the fiber core.

S. Todoroki, *Fiber Fuse*, NIMS Monographs, DOI: 10.1007/978-4-431-54577-4_5,
© National Institute for Materials Science, Japan. Published by Springer Japan 2014

This new model also provides practical approaches for preventing fiber fuse problems. The delayed response of the glass melt enables us to realize fiber fuse termination based on a steep MFD modulation. Colored fiber coatings and bare segments sometimes activate the plasma via self-pumping. The idea of fiber fuse propagation modes in typical single-mode fibers help us to postulate fiber fuse behavior in fibers with a more advanced and complex structure including holey fibers and multicore fibers.

Box 5.1 An anecdote of the author's early stage of fiber fuse research

Originally written in Japanese for domestic news letters.

- Two serendipitous episodes – How I embarked on fiber fuse research (2007)
  http://pubman.nims.go.jp/pubman/item/escidoc:33124

Appendix A
Comparison with Bulk Silica Glass Modification by Continuous-Wave Laser

Extremes meet.

In 2006, an interesting phenomenon was reported by researchers in precision engineering, in which bulk silica glass is modified with a continuous-wave laser beam focused on a thin copper foil that covers the backside of the glass [10] (see Fig. A.1). The modification starts with heat-induced generation of plasma (or optical discharge) on the foil. Then, it moves to the light source, that was clearly recorded by high-speed photography [2, 4, 6]. Finally, it disappears when the power supply is exhausted due to the laser beam defocusing at the plasma [2]. This resembles the fiber fuse initiation, propagation, and termination processes discussed in Chap. 1.

Sometimes, the modified area contains hollow voids [2, 4, 6]. Figure A.2 describes one example of this modification process reported in Hidai et al. [4] where a void train of ~5.5 mm was left. The void shape varies with the depth from the copper foil; (a) periodic bullet-like voids near the foil, (b) quasi-periodic voids with a smaller interval and irregular shape in the middle of the train, and (c) a long void (>400 μm) with a slight periodic necking at the end of the train. This trend is driven by the pump power density reduction at the plasma front. It also reminds us of the following fiber fuse behaviors; the reduction of periodic void interval with the pump power in the cylindrical mode as shown in Fig. 3.1a and the long void generation in response to a rapid power reduction as shown in Fig. 4.5d.

Hidai et al. [4] analyzed the modified area through fictive temperature determined by Raman and IR spectroscopy, Vickers hardness and etching rate. They revealed that the area consists of two layers. The inner zone is mainly modified by laser heating and the outer by tensile stress caused by the densification of the inner area. It is reasonable to expect that this structure is also left after a passage of a fiber fuse.

In this connection, it should be noted that the CW-LBI method triggers another interesting phenomenon where a metal particle in a glass block is pumped to move, radiate light/heat, and modify the surrounding [3]. One example is shown in Fig. A.3.

S. Todoroki, *Fiber Fuse*, NIMS Monographs, DOI: 10.1007/978-4-431-54577-4,
© National Institute for Materials Science, Japan. Published by Springer Japan 2014

Fig. A.1 Experimental setup for continuous-wave laser backside irradiation (CW-LBI)

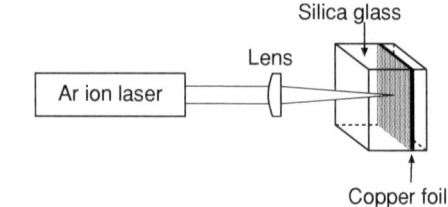

Fig. A.2 Modification process of bulk silica glass by CW-LBI; (*top*) initiation, (*middle*) propagation, and termination, and (*bottom*) the positions of damage photographs are shown in Fig. 3 in Hidai et al. [4]

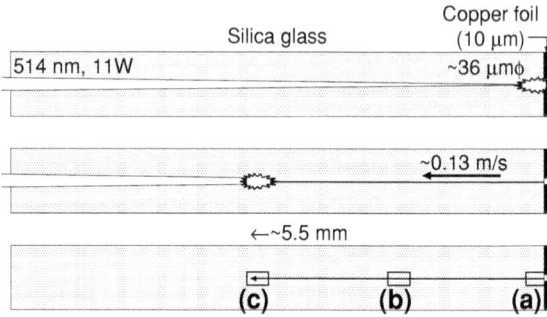

Fig. A.3 Migration process of platinum particle (diameter:~5 μm) in borosilicate glass by CW-LBI [3]; (*top*) initiation, (*bottom*) propagation and termination

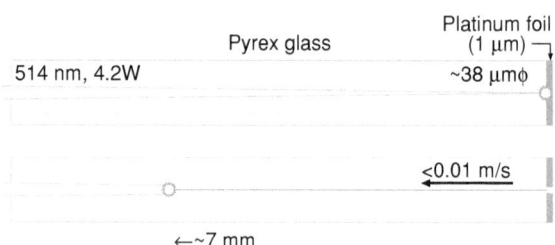

A metal particle is introduced from the platinum foil on the backside of a borosilicate glass piece and moves until the power supply is exhausted due to defocusing. The driving force of particle movement is considered to be the gradient of surface tension that is introduced by the poor thermal conductivity of the metal, i.e., a thermal gradient between the fore and rear of the particle [5]. Compared with the case of copper foil on silica glass shown in Fig. A.2, the propagation speed is slower by an order of magnitude at least.

Appendix B
Fiber Fuse in Materials Other than Silica Glass

Never judge by appearances.

B.1 Soft Glass Fibers

As described in Sect. 1.2, a distraction wave without plasma is known to occur in chalcogenide and fluoride fibers and to decompose the entire cross section of the fiber [1].

B.2 Polymer Optical Fiber

In January 2014, it was reported that a fiber fuse phenomenon occurs in a polymer optical fiber (POF) [8]. Its macroscopic appearance is in perfect analogy to what we have seen in silica glass fibers except its slow propagation speed, ~ 0.02 m/s. A bright spot runs through a perfluorinated graded-index POF having a core of 50 μm diameter pumped with <0.2 W, 1,546-nm light. The most striking feature is found in its microscopic behavior. The damage left behind the spot looks like a black wavy line oscillated at the wavelength intrinsic to its graded-index profile as shown in Fig. B.1. A unique termination method is presented thanks to the high elasticity of the polymer. A fiber fuse stops at a point pressed with an outer metal attachment. These facts suggest that its propagation mechanism is completely different from that in silica glass fibers.

S. Todoroki, *Fiber Fuse*, NIMS Monographs, DOI: 10.1007/978-4-431-54577-4,
© National Institute for Materials Science, Japan. Published by Springer Japan 2014

Fig. B.1 An oscillatory curve of fiber fuse damage in a polymer optical fiber with graded-index profile. The outer diameter is 750 μm. Courtesy of Dr. Yosuke Mizuno, Tokyo Institute of Technology

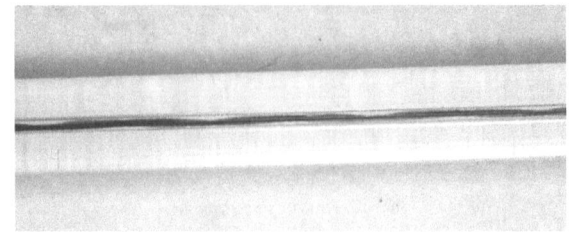

Box B.1 Video providing macroscopic view of polymer optical fiber fuse.

• Fiber fuse propagating along a polymer optical fiber at an extremely slow speed (~ 0.02 m/s).

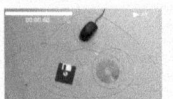   http://www.youtube.com/watch?v=t0k_B6EOQhg
Duration: 1:11
Fiber: Perfluorinated graded-index polymer fiber
Pump laser: 1546 nm, 75 mW
Speed: 30 fps

B.3 Yb-Doped Bismuthate Glass Waveguide

Fiber fuse phenomenon is also observed in active devices made of silica glass fibers including laser emitters [11] and optical amplifiers [9]. In addition, it was recently reported that an active waveguide laser device fabricated in a Yb-doped bismuthate glass was destroyed by this phenomenon leaving a bullet-shaped void train [7].

References

1. E.M. Dianov, I.A. Bufetov, A.A. Frolov, V.M. Mashinskii, V.G. Plotnichenko, M.F. Churbanov, G.E. Snopatin, Catastrophic destruction of fluoride and chalcogenide optical fibers. Electron. Lett. **38**(15), 783–784 (2002). doi:10.1049/el:20020539
2. H. Hidai, M. Yoshioka, K. Hiromatsu, H. Tokura, Glass modification by continuous-wave laser backside irradiation (CW-LBI). Appl. Phys. A **96**(4), 869–872 (2009). doi:10.1007/s00339-009-5324-x
3. H. Hidai, T. Yamazaki, S. Itoh, K. Hiromatsu, H. Tokura, Metal particle manipulation by laser irradiation in borosilicate glass. Opt. Express **18**(19), 20313–20320 (2010). doi:10.1364/OE.18.020313
4. H. Hidai, M. Yoshioka, K. Hiromatsu, H. Tokura, Structural changes in silica glass by continuous-wave laser backside irradiation. J. American Ceram. Soc. **93**(6), 1597–1601 (2010). doi:10.1111/j.1551-2916.2010.03615.x

5. H. Hidai, M. Matsushita, S. Matsusaka, A. Chiba, N. Morita, Moving force of metal particle migration induced by laser irradiation in borosilicate glass. Opt. Express **21**(16), 18955–18962 (2013). doi:10.1364/OE.21.018955

6. S. Itoh, H. Hidai, H. Tokura, Experimental and numerical study of mechanism of glass modification process by continuous-wave laser backside irradiation (CW-LBI). Appl. Phys. A **112**(4), 1043–1049 (2013). doi:10.1007/s00339-012-7477-2

7. R. Mary, D. Choudhury, A.K. Kar, Applications of fiber lasers for the development of compact photonic devices. IEEE J. Sel. Top. Quantum Electron. **20**(5), 902513 (2014). doi:10.1109/JSTQE.2014.2301136

8. Y. Mizuno, N. Hayashi, H. Tanaka, K. Nakamura, S. Todoroki, Observation of polymer optical fiber fuse. Appl. Phys. Lett. **104**(4), 043302 (2014). doi:10.1063/1.4863413

9. J. Wang, S. Gray, D. Walton, L. Zenteno, Fiber fuse in high-power optical fiber, *in Passive Components and Fiber-based Devices V* ed. by M.J. Li, P. Shum, I.H. White, X. Wu. *SPIE Proceedings*, vol. 7134, pp. 71342E–1-9. SPIE, Hangzhou, China (2008). doi:10.1117/12. 803114

10. M. Yoshioka, H. Hidai, H. Tokura, CW-laser induced modification in glasses by laser backside irradiation (LBI), in *Photon Processing in Microelectronics and Photonics V* ed. by T. Okada, C.B. Arnold, M. Meunier, A.S. Holmes, D.B. Geohegan, F. Träger, J.J. Dubowski. *SPIE Proceedings*, vol. 6106, pp. 61060Y–1-8. SPIE (2006). doi:10.1117/12.645829

11. H. Zhang, P. Zhou, X. Wang, H. Xiao, X. Xu, Fiber fuse effect in high-power double-clad fiber laser. In: Conference on lasers and electro-optics Pacific Rim (CLEO-PR) (2013), Paper WPD-4. doi:10.1109/CLEOPR.2013.6600547

Index

S. Todoroki, *Fiber Fuse*, NIMS Monographs, DOI: 10.1007/978-4-431-54577-4,
© National Institute for Materials Science, Japan. Published by Springer Japan 2014